NUCLEAR WEAPONS

1945 onwards (strategic and tactical delivery systems)

'I know not with what weapons World War Three will be fought, but World War Four will be fought with sticks and stones.'

Albert Einstein

'…how well we had done our dreadful work.'

Jacob Bronowski

First published in June 2017
Reprinted in November 2018, May 2022, August 2023 and October 2024

A catalogue record for this book is available from the British Library.

ISBN 978 1 78521 139 3

Library of Congress control no. 2017932254

Published by Haynes Group Limited,
Sparkford, Yeovil, Somerset BA22 7JJ, UK.
Tel: 01963 440635
Int. tel: +44 1963 440635
Website: www.haynes.com

Haynes North America Inc.,
2801 Townsgate Road, Suite 340
Thousand Oaks, CA 91361

Printed in India.

COVER IMAGE:
Britain's nuclear deterrent is due for an upgrade to give the Trident SLBM a new class of submarine. *(USN)*

NUCLEAR WEAPONS

1945 onwards (strategic and tactical delivery systems)

Operations Manual

An insight into the science, design and engineering of
tactical and strategic nuclear delivery systems

David Baker

Contents

ABOVE RIGHT The Ivy Mike shot on 1 November 1952 had a yield of 10.4MT, the world's first thermonuclear explosion. *(LANL)*

OPPOSITE On 8 June 1958 the Hardtack Umbrella shot detonated an 8KT device 46m (150ft) below the surface at the bottom of the Pokon lagoon, Eniwetok, to test the effect of nuclear weapons on ships. *(LANL)*

Introduction

Since their first use against the Japanese cities of Hiroshima and Nagasaki, nuclear weapons have frequently been at the forefront of news, public interest and political attention. This book explains the way those weapons were developed and a little bit of the science and engineering behind their design and assembly. It also describes the development of nuclear weapons in the countries and states that have them and those who aspire to obtain them.

The book does not judge either the moral or political issues, nor does it discuss the way these weapons have influenced military thinking about tactical and strategic warfare – there are plenty of such books out there already. However, what I have tried to do is to show the way nuclear weapons have, to some extent, become "normalised" in the lexicon of military language and in the files of active war-plans through their inclusion in force inventories.

It could be said that nuclear weapons have dominated political, diplomatic and military war planning since the end of the Second World War, and to some extent they continue to do so today. Yet there are only nine nuclear powers among the more than 200 sovereign states in

the world, all but North Korea being capable of delivering nuclear warheads or bombs to targets outside their borders.

National defence has very little to do with the possession of nuclear weapons and they can sometimes be an impediment to diplomacy or political solutions, yet those who have them are loath to give them up. Paradoxically, in addition to biological and chemical weapons, they are classed as one of the three weapons of indiscriminate mass destruction – WMD – outlawed in international law.

In overall balance, the contribution made by the numerically minor nuclear powers (all but the US and Russia) is accounting for an increasingly high proportion of the global inventory, this being due to the significant reduction in the inventories of the US and Russian forces and the modest growth in nuclear weapons by the other six (or seven when North Korea puts bombs on missiles).

In many respects, the possession of nuclear weapons traps nations in the past, a situation similar to the early days of the space programme at the heart of the Cold War; they are possessed as technological virility symbols, displays of absolute power but difficult to integrate within

BELOW Britain's HMS Vanguard off Cape Canaveral, Florida, in April 1994, epitomises the UK's nuclear deterrent. *(RN)*

a defensive structure because they cannot be used. This because they are offensive weapons only of real use for the pre-emptive destruction of an enemy, either on the battlefield or to destroy cities and vast urban spaces.

Arguably their most dangerous application is not in war at all. In 1969 President Nixon and his national security adviser Henry Kissinger set up the "madman plan" which was to send a veiled message to the Soviet Union that unless Russia forced North Vietnam to peace talks he would unleash nuclear war – implying that the President was at his wits' end and was out of control.

Invoking unjustified fear as a fabricated game of international poker, on 10 October Nixon sent nuclear armed B-52s racing for the Russian border as if to reinforce his determination. The Russians held their nerve and Nixon failed to repeat his dangerous game of brinkmanship. But on another occasion he put the entire nuclear arsenal on full alert – as a reminder to the Russians. On other occasions, faulty electronic early-warning systems tipped the world to the very edge of an all-out nuclear strike while accidents and mishaps have come close to unintentional detonation.

This then is the story of the nuclear weapon – hard to develop, difficult to give up, almost impossible to use. Nuclear forces are justified on the basis that their use would invoke a response which would bring unacceptable destruction in return, therein becoming a deterrent. That is the logic of mutually assured destruction (MAD). There are those who seek to silence debate and stifle questions, fearful that the answers may invalidate their assumptions. The aim of this book is to inform and to leave those judgements to the reader.

David Baker
East Sussex
June 2017

ABOVE The consequences of even a limited war are open-ended and incalculable, potentially threatening to human life on Earth. *(Via David Baker)*

Acknowledgements

I would like to thank several people who have influenced my work on nuclear weapons theory and analysis including giants at whose feet I had the opportunity to learn much, specifically Hermann Kahn, Edward Teller, Lawrence Freedman and many others, personnel who have worked on the development of nuclear weapons, members of the armed forces who have worked with them and personnel who have serviced them.

Specifically, my thanks to Haynes for asking me to write this book, to Steve Rendle for managing the project, to Ian Heath for editing the text and to James Robertson for putting it all together. I would also like to thank that erudite aerospace historian Chris Gibson for helping with some of the illustrations. Last, but not least, thanks to my dear wife Ann for her continued support and sustained encouragement.

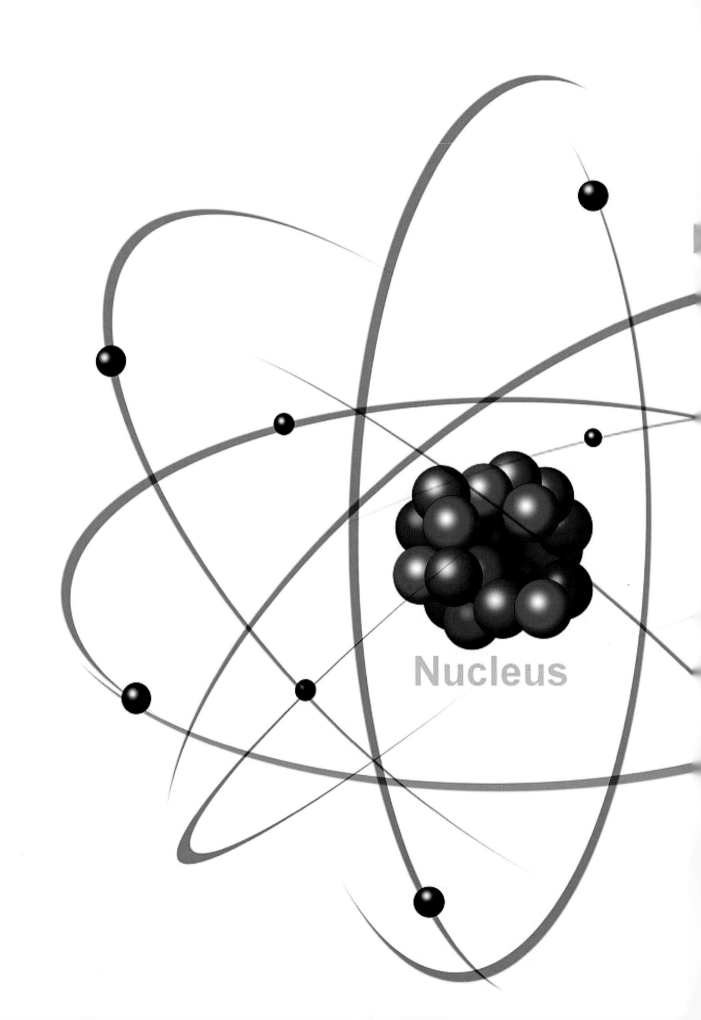

Nucleus

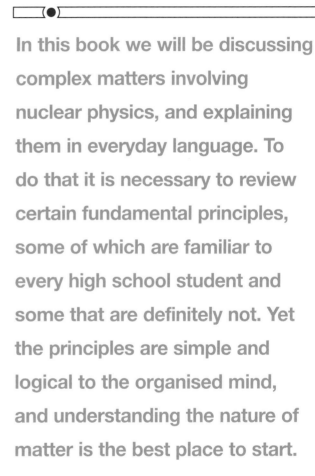

Chapter One

The new physics

In this book we will be discussing complex matters involving nuclear physics, and explaining them in everyday language. To do that it is necessary to review certain fundamental principles, some of which are familiar to every high school student and some that are definitely not. Yet the principles are simple and logical to the organised mind, and understanding the nature of matter is the best place to start.

OPPOSITE The simple model of the atom taught to school children is an over-simplification but the basic starting principle is relevant, the nucleus surrounded by negatively charged electrons, the number proportional to the positively charged protons. *(Jafari)*

ABOVE Energy liberated by the manipulation of atoms followed centuries of molecular manipulation by chemists to support the industrial evolution. But the energy liberated through nuclear fission, and later fusion, was greater by several orders of magnitude. Dubbed the 'runaway bomb', America's Castle Romeo test of 27 March 1954 produced a yield of 11MT versus the expected 4MT. *(LANL)*

RIGHT A Russian chemist born in Siberia to a successful middle-class family, Dmitri Mendeleev (1834–1907) taught in St Petersburg before taking on a professorship at the Technological Institute. In 1869 he set out the known elements in an elegant, yet logical, set of linked connections that would account for the fundamental properties of matter – the periodic table. *(David Baker)*

Scientists have known for some considerable time that all matter is made up of atoms, of which there are more than 100 different types, each one defining a specific element such as carbon, iron, zinc, platinum and uranium, to name but a few. Elements are the fundamental building blocks of chemistry and of chemical substances that consist of individual atoms. Individual atoms can be made to join together to form compounds, the smallest of which are molecules.

Working at the St Petersburg State University in Russia, Dmitri Mendeleev was fascinated by chemical elements, and in the early 1860s he devised a way to categorise them and to lay them out in a table in order of their composition and properties. Because of this he was able to extrapolate beyond the then 62 known elements and predict the existence of those not yet discovered. This work resulted in his 'periodic table' of 1869, since which date all of chemistry has flowed universally around the world.

The periodic table came about as a logical progression after Mendeleev had first arranged the elements in a long line in order of their atomic mass. He noticed that if this was cut up into strips and arranged in separate rows to make a table, each column contained elements with similar properties. For instance, sodium, lithium and potassium all react violently with water, remain solids at room temperature and tarnish readily in air. He called all 18 of these columns 'groups', and the repetition of properties in rows he named 'periodic'.

Mendeleev was working to define the separate elements according to chemical properties but in doing so he was laying the groundwork for other ways in which the atom can be put to work, ways that form the fundamental basis of nuclear physics. For it is through the universal application of the periodic table that the separate fields of chemistry and physics join. Gradually, through Mendeleev's table, other scientists began to play with the possibilities, placing British physicist Henry Moseley on the path to proving, in 1913, that the true measure of an individual atom is the measure of the charge on its nucleus.

It was only through rigorous discipline that Mendeleev proved the validity of his periodic table. It worried him that when he first placed arsenic in group 13, period 4, it did not seem to

| Group→ | 1 | 2 | 3 | 4 | 5 | 6 | 7 | 8 | 9 | 10 | 11 | 12 | 13 | 14 | 15 | 16 | 17 | 18 |
↓Period																		
1	1 H																	2 He
2	3 Li	4 Be											5 B	6 C	7 N	8 O	9 F	10 Ne
3	11 Na	12 Mg											13 Al	14 Si	15 P	16 S	17 Cl	18 Ar
4	19 K	20 Ca	21 Sc	22 Ti	23 V	24 Cr	25 Mn	26 Fe	27 Co	28 Ni	29 Cu	30 Zn	31 Ga	32 Ge	33 As	34 Se	35 Br	36 Kr
5	37 Rb	38 Sr	39 Y	40 Zr	41 Nb	42 Mo	43 Tc	44 Ru	45 Rh	46 Pd	47 Ag	48 Cd	49 In	50 Sn	51 Sb	52 Te	53 I	54 Xe
6	55 Cs	56 Ba	57 La	* 72 Hf	73 Ta	74 W	75 Re	76 Os	77 Ir	78 Pt	79 Au	80 Hg	81 Tl	82 Pb	83 Bi	84 Po	85 At	86 Rn
7	87 Fr	88 Ra	89 Ac	** 104 Rf	105 Db	106 Sg	107 Bh	108 Hs	109 Mt	110 Ds	111 Rg	112 Cn	113 Nh	114 Fl	115 Mc	116 Lv	117 Ts	118 Og

		58 Ce	59 Pr	60 Nd	61 Pm	62 Sm	63 Eu	64 Gd	65 Tb	66 Dy	67 Ho	68 Er	69 Tm	70 Yb	71 Lu
*		58 Ce	59 Pr	60 Nd	61 Pm	62 Sm	63 Eu	64 Gd	65 Tb	66 Dy	67 Ho	68 Er	69 Tm	70 Yb	71 Lu
**		90 Th	91 Pa	92 U	93 Np	94 Pu	95 Am	96 Cm	97 Bk	98 Cf	99 Es	100 Fm	101 Md	102 No	103 Lr

fit, having more in common with the properties of group 15. That left two vacant slots, which he presumed would be filled with elements yet to be discovered. And that was another beautiful aspect to the way he worked: he always assumed that he could never know the whole picture and only worked to make sense of what he had. So he placed arsenic in group 15, which left places in groups 13 and 14 filled later by the discovery of gallium and germanium!

At the time, Mendeleev believed that the atom was the smallest unit of matter, and that its mass – its 'atomic number' – defined the nature of the element. As we now know, it is a much more complex picture and becomes increasingly so as additional knowledge is bolted on to the existing base. The atom contains a nucleus of positively charged protons and neutrons, which possess no charge. In simplified explanation this nucleus of protons and neutrons is surrounded by orbiting electrons, each one of which balances the opposing charge of the similar number of protons. But the path to that knowledge was long and tortuous.

The development of the science of nuclear physics occurred through a series of experiments, discoveries and developments in basic research across several decades. The electron was discovered by J.J. Thomson – for which he won the Nobel Prize in 1908 – while carrying out experiments in which air was pumped out of a tube and a high voltage was applied between two electrodes at either end. The resultant stream of particles from the negatively charged electrode, the cathode, across to the positively charged anode was called a 'ray' and the whole device was known as a cathode ray tube.

ABOVE The periodic table is the foundation of chemistry, and the guidebook to the elementary basis of nuclear physics as well. Arranged by atomic number, which defines the number of protons in the nucleus, arranged in nine rows, the elements are grouped in 18 columns to show similar chemistry. *(David Baker)*

LEFT Henry Mosely (1887–1915) is best known for his work establishing the atomic number and the charge on the nucleus as well as X-ray emissions from elements. After turning down a position with Oxford University to enrol in the Army at the outbreak of war in 1914 he was eventually sent to Gallipoli, where he was killed on 15 August 1915 at the age of 27. *(Via David Baker)*

RIGHT Joseph John Thomson (1856–1940) is best known for his discovery and description of the electron, a neutrally charged particle to balance the charge on the atom from the positively charged proton. His fundamental work established the science to develop the cathode ray tube, although his contributions led to a Nobel Prize in 1906 for his description of the conductivity of electricity in gases. *(Via David Baker)*

ABOVE RIGHT Attributed with making the greatest contribution to nuclear physics, Ernest Rutherford (1871–1937) discovered radioactive half-life, which began the study of the transmutation of elements from one into another through natural decay. Born in New Zealand, he became director of the Cavendish Laboratory at the University of Cambridge in 1919 and is arguably the most important figure in the physics of atoms and matter, inspiring an entire generation of nuclear physicists. *(Via David Baker)*

BELOW Rutherford's laboratory was awash with experimentation born from new ideas and novel research. He was one of the first 'outsiders' allowed to carry out research at Cambridge from 1871, when it was almost impossible to gain a position from outside the enclosed walls of that academic fraternity. *(Cavendish Laboratory)*

But Thomson did not discover the absolute value of the charge of the electron, that being determined by Robert Millikan in 1909 when he allowed a fine spray of oil to settle through a hole into a chamber, where he observed its reaction to electrically charged plates at top and bottom. By introducing a series of X-rays, which can cause discharge on contact with matter, Millikan observed the rate of fall between objects with negative charge compared to those with a positive charge, the former stopping falling or falling more slowly than those with a positive charge at a rate depending on the number of charges on them.

Working at the University of Chicago, Millikan observed the reaction of the droplets with varying degrees of charge or voltage on them and calculated that a single electron carries a charge of -1.60×10^{-19}C (coulombs). From this he discovered that the mass of an electron is 9.1×10^{-28}gm and that the negatively charged particles have a charge/mass ratio of -1.76×10^{8}C/gm. The result of the work originally carried out by Thomson posed the question regarding the location of the positive charge necessary to create an electrically neutral atom.

The New Zealand physicist Ernest Rutherford received the Nobel Prize in 1908 for his work on the discovery of radiation by working with Hans Geiger and demonstrating that alpha particles were ionised helium nuclei. Together with

Ernest Marsden, in 1911 Rutherford showed through a famous gold foil experiment that in all probability the nucleus was very small and that it was positively charged and orbited by negatively charged electrons. Just before he left Manchester University in 1919 to take charge of the Cavendish Laboratory at Cambridge University, Rutherford became the first person to effect a nuclear reaction by determining that the conversion of hydrogen into oxygen through $^{14}N + \square \rightarrow \, ^{17}O +$ (positively charged particle). The following year, Rutherford named that positively charged particle the proton.

By 1920 physicists knew that the atom was largely empty space, that very powerful forces exerted control of radioactive decay, that the mass of the atom is concentrated in a very small core surrounded by electrons, their abundance determined by the number of protons, and that this was why the atom appeared electrically neutral. But the model was not complete, because the mass of the proton was only half that of the atom, and Rutherford was convinced that there was another particle, of approximately the mass of the proton but without any electrical charge.

A wide-ranging scientific debate ensued, and in 1930 W. Bothe and H. Becker discovered that an electrically neutral radiation would be released when beryllium was bombarded with alpha particles. The conclusion was that photons with high-energy gamma rays were being released. Two years later, Irène and Frédérick Joliot-Curie showed that this

ABOVE Like most good science, inspiration comes from association, and Rutherford worked with J.J. Thomson, who recommended him for a post at McGill University in Montreal, Canada, in which establishment he had his own laboratory, seen here. In 1920 Rutherford defined the characteristics of the particle he named the proton. *(Via David Baker)*

BELOW LEFT The American physicist Robert Andrews Millikan (1868–1953) was the first to measure the precise charge of the electron and from that to determine its mass. *(Via David Baker)*

BELOW Working with his wife Irène (1897–1956), Frédérick Joliot-Curie (1900–58) focused his work on the structure of atoms and on the nature of nuclei, which in turn encouraged Chadwick to discover the neutron in 1932. Irène was the daughter of the renowned physicist Marie Curie (1867–1934), born in Warsaw when it was part of Russia, who set up methods of isolating radioactive isotopes and discovered polonium and radium. *(Via David Baker)*

could compensate for the proton's charges in the nucleus and serve as a binding unit within the nucleus. An associate of Rutherford, Chadwick received the Nobel Prize in 1935.

From these came the derivations and calculations that determined that a proton (p^+) has a mass of 1.672×10^{-27}kg, that the neutron (n^0) has a mass of 1.674×10^{-27}kg and that the electron (e^-) has a mass of 9.1×10^{-31}kg. Both electron and proton have a charge of -1.60×10^{-19}C. Thus were the known properties of the atom in place for the next evolution in theoretical physics – the equations that would allow the enormous quantities of energy within the atom to be liberated.

Isotopes

By the late 1930s protons and neutrons were known to be much larger than the electron and could be interchangeable. For instance, a proton can become a neutron if it absorbs an electron, where the positive charge cancels the negative one. A neutron can become a proton if it expels a unit of electrical charge and ejects an electron. Each element has a characteristic number of protons in the nucleus, balanced by a similar number of electrons, and this determines the nature of the element. In chemical reactions the neutrons have little effect, but in nuclear reactions they are very important.

could eject protons when hit by paraffin or compounds containing hydrogen. But this begged the question as to how a photon, without mass, could eject protons that are 1,836 times the mass of electrons?

In 1932 James Chadwick carried out the same experiment but used several different targets and, through analysis of the different energies of the bombarded materials, discovered the existence of the neutron. More than ten years earlier, while working with Niels Bohr, Rutherford had postulated the existence of a second particle in the nucleus, one which

Usually the nucleus has the same number of neutrons and protons, the latter being fixed as a determinant of the element. But the neutron count can vary and this determines the isotope of the element. For instance, carbon has six neutrons and six protons and is referred to as carbon-12, but it can acquire a second neutron to become carbon-13 or a fourth to become carbon-14. But they are all carbon and therefore have the same atomic number, 6.

There are cases where isotopes are so different that they get a different name. Hydrogen has no neutron, only one proton balanced by one electron, and is therefore the lightest element. But a percentage of natural hydrogen can possess a neutron (deuterium) or two (tritium). Deuterium can combine with water to make heavy water, so called because the mass of the hydrogen 'isotope' is greater than in natural hydrogen.

Hydrogen (^{1}H) accounts for 75% of all the mass in the universe, but deuterium (^{2}H) was created in the Big Bang that formed the physical universe, while tritium (^{3}H) is radioactive and decays, with a half-life of 12.32 years.

Radioactive decay allows one element to change into another through either alpha-decay or beta-decay. In alpha-decay the nucleus loses a binding pair of two protons and two neutrons, which lowers the mass number by four units but the atomic mass by only two (through the departing protons). Thus will an element transform itself into another element, defined by the lower number of protons. For instance, uranium 238 has 92 protons and 146 neutrons (92+146=238). But when it decays by losing a binding pair it becomes thorium 234 with 90 protons and 144 neutrons.

Beta-decay occurs when one neutron in the nucleus becomes a proton by ejecting a single unit of negative charge or an electron. The mass number remains the same because there is no difference in the total number of protons and neutrons, but the atomic number is raised by 1 because of the additional proton. In this way, for example, sodium 24 becomes magnesium 24 through beta-decay, increasing the atomic number but retaining the same mass number, 24. The atomic number has increased from 11, unique to sodium, to 12, unique to magnesium.

In another form of transformation, a stray

neutron can collide with a nucleus, knock out a proton and take its place. But there is no change in the mass number, only the atomic number has decreased by 1 because of the loss of a proton. Yet another transformation can occur when a proton turns into a neutron by acquiring an electron, neutralising its charge. This lowers the atomic number by 1 but the mass number remains the same. In this way potassium 40, which has the atomic number 19, decays out to argon 40, with atomic number 18.

The rate of decay is measured by half-life, which we referred to earlier, in effect an inverse

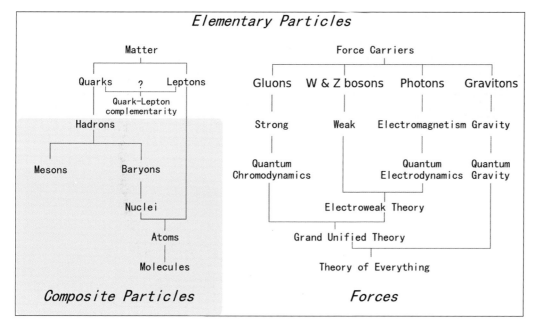

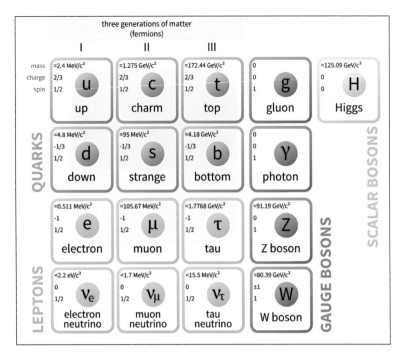

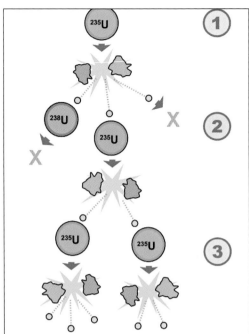

ABOVE The standard model of elementary particles accepts that there are some with defined substructures and it is through these that the connection with the quantum world is established, as well as between the particles themselves and their carrier-forces. *(Via David Baker)*

ABOVE RIGHT A nuclear fission chain reaction. 1, A uranium-235 atom absorbs a neutron and fissions into two new atoms (fission fragments), releasing three new neutrons and some binding energy; 2, one of those neutrons is absorbed by an atom of uranium-238 and does not continue the reaction. Another neutron is simply lost and does not collide with anything, also not continuing the reaction. However, one neutron does collide with an atom of uranium-235, which then fissions and releases two neutrons and some binding energy; 3, both neutrons collide with uranium-235 atoms, each of which fissions and releases between one and three neutrons, which can then continue the reaction. *(David Baker)*

BELOW Naturally occurring isotopes of hydrogen (left to right): protium, with one proton and one electron; deuterium, with an additional neutron; tritium, with two neutrons. *(Balajijagadesh)*

asymptotic rate in which half the isotope will decay in a given time, followed by half of the remainder in the same period again, and half of that in a further similar duration. It never reaches zero. For instance, the half-life of strontium 90 is 28 years at the end of which half the initial quantity will have decayed into yttrium 90, which will eventually change into zirconium 90. The half of the strontium 90 remaining will take a further 28 years to reduce by half again – and so on ad infinitum.

The exact rate of decay to an absolute nth decimal place is unpredictable because this, like other aspects of atomic physics, is governed by quantum mechanics, which dictates that the presence of a particle will never be predictable but only statistically probable over a protracted period. Consequently the longer the observation, the more accurate the prediction – which, turned on its head, means that the presence of the particle will, like its characteristics, only reveal itself if asked the right question. Physicists know that a particle that has spin, charge and mass will only declare the value of the parameter when that particular value is interrogated.

But there is room for manoeuvre in controlling what happens in the confines of an atom with orbiting electrons – a lot of it. As defined by the 'virtual' orbit of the electrons around the nucleus, an atom is largely empty

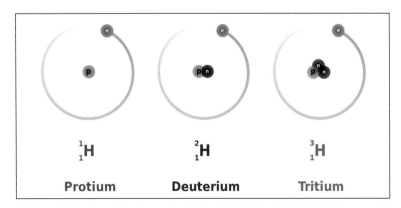

^1_1H ^2_1H ^3_1H

Protium   **Deuterium**   **Tritium**

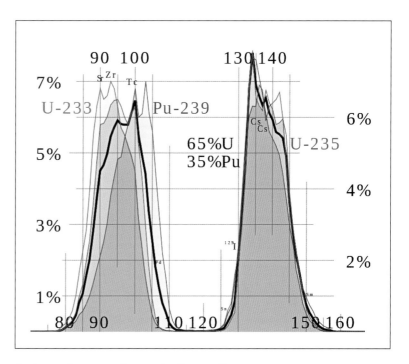

RIGHT Yields by mass for thermal fission of uranium-235 and plutonium-239, which explains why these two atoms liberate the greatest amount of energy in a fission process. (Via David Baker)

space. When substances react chemically they are sharing electrons, and that is a product of quantum reactions.

Determined by the number of protons, the number of electrons in an atom are organised into various levels outward from the nucleus, in shells that hold specific quantities. The innermost shell carries two electrons, the next shell can have eight, the third shell up to 18 and so on, the total number of electrons equating to the number of protons in the nucleus. The electrons in the outermost shell, determined by the chemical properties of that atom, are in what is known as the valence shell and it is across the valence boundaries that groups of atoms link to form a compound.

Within the structure and forces at work in the atom are mechanisms for liberating a very large amount of energy, and it is this factor which opens the possibility of using the atom to create a very powerful bomb, roughly one million times more energy than could be liberated with an equivalent chemical reaction. To understand why, it helps to appreciate the forces inherent within the structure and nature of the atom and its nucleus.

There are four known forces: gravitational, weak nuclear, electromagnetic, and strong nuclear. Until the unification of electrical and magnetic forces there were five, and scientists are now seeking a way to explain all four through a unifying theory. Current models integrate the electromagnetic and the weak nuclear into what is called the 'electroweak' force. However, to appreciate the advantage that will soon become apparent, here we will continue to regard them as four separate forces.

Gravity is responsible for attraction at a molecular level and is the reason why the universe can accrete gases into stellar systems, debris into planets and fix humans to the surface of the Earth. It is the weakest of these forces but has infinite range, decreasing on the inverse square law, but it is the only force which is cumulative and multiplies on a linear scale:

the larger the mass, the greater the gravity (which determines the 'weight' of the mass).

The weak nuclear force is responsible for some nuclear phenomena such as radioactive decay, and is 10^{25} times the strength of gravity. It is an essential force in nuclear weapons. Electromagnetism is the force that acts between electrically charged particles. While it is infinite in range, similarly declining on the inverse square law, it is 10^{36} times the strength of gravitational force.

The strong nuclear force is the most complex of all and is basically regarded as the binding energy that holds the particles of the nucleus together, being 10^{38} times the strength of gravity. Yet inside the nucleus it has an effective radius only as great as the diameter of a nucleic particle itself. This is the second important force for nuclear weapons design.

LEFT A billet of highly enriched uranium of the type that would be employed in a fission bomb. (Via David Baker)

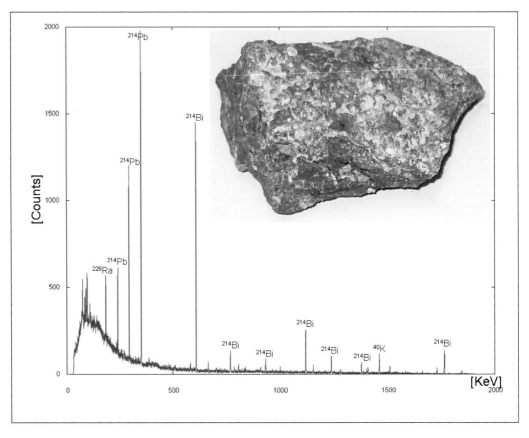

Gaining energy

Any nuclear reaction in which there is a net decrease of mass, in which the total mass of the products is less than that of the interacting particles, will be accompanied by the liberation of energy. Two types are possible: fission, the splitting of a heavy nucleus into a pair of lighter nuclei; and fusion, the combination of two very light nuclei to form a heavier one. The underlying principle in both cases is that the average net energy attraction between the nucleons (protons and neutrons) is smaller in the initial nucleus than it is in the products of the reaction.

The magnitude of this liberated, or binding, energy can be easily calculated from the masses of the various particles. Determination of the mass defect can be calculated if A is the mass number and Z is the atomic number, so that the nucleus contains Z protons and A-Z neutrons. If m_p is the mass of a proton, m_n is the mass of a neutron, and M is the actual mass of the nucleus, then the mass defect (MD) of a particular isotope is defined by: MD = [Zm_p + (A-Z)m_n] – M. The mass defect, which can be taken as the decrease of mass that would result if Z protons and A-Z neutrons were combined to form a given nucleus, is a measure of the binding energy of that nucleus.

If the various masses are expressed in atomic mass units (amu), where m_p is 1.00813 and m_n is 1.00897, multiplication of MD by 931 provides the binding energy (BE) in millions of electron volts (MeV). This factor is based on Einstein's mass energy equation using the appropriate units of mass (amu) and energy (MeV). Consequently: BE = 931 {[Zm_p + (A-Z)m_n]-M} MeV. This provides the value of the energy released in the formation of a nucleus by a combination of the appropriate protons and neutrons, or, alternatively, the energy that would be necessary to break up the nucleus into its constituent protons and neutrons.

An interesting observation for the nuclear physicist is that the mean binding energy per nucleon is less for the lightest and for the heaviest nuclei than it is for those of intermediate mass. It is this that accounts for the liberation of energy accompanying either the fission of heavy nuclei or the fusion of light nuclei. Uranium-235 undergoes fission in about

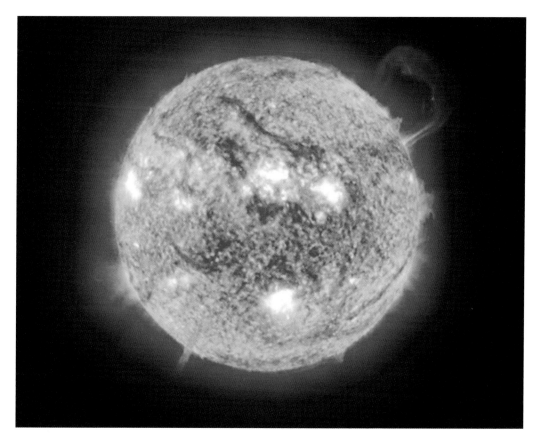

30 different ways, but the mass number of fission products lies within the mass range of 80 to 150, and the average binding energy per nucleon in this range is about 8.4MeV.

However, in the original uranium-235 the binding energy per nucleon is about 7.5MeV, which is the total amount of energy required to break up the uranium-235 nucleus into its 235 constituent nucleons. On subtraction of the higher value from the lower: uranium-235 → fission products + (235 x 0.9) MeV, showing that the fission of U^{235} is accompanied by the release of 235 x 0.9MeV, which is about 202.5MeV of energy.

While fission of heavy nuclei can be brought about in a number of different ways, there is only one that is of importance for the practical release of energy and that is through the use of neutrons. Because the fission process itself is accompanied by the liberation of neutrons, a chain reaction ensues to ensure sustained reaction. Only three isotopes are practical for the fusion process, U-233 (bred from thorium-232), U-235 and Pu-239. These substances are radioactive and have relatively long half-lives, which means they are

moderately stable and will undergo fission by the capture of neutrons of all energies, either fast (high energy) or slow (low energy).

Uranium-238 accounts for 99.3% of the

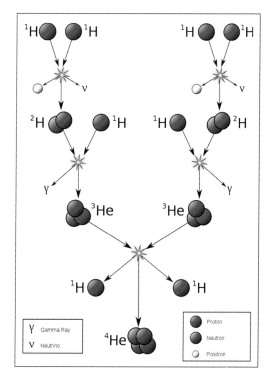

LEFT The fundamental stages by which nuclear fusion on the proton-proton chain combined hydrogen nuclei to produce helium, the same process as that in our Sun. (Via David Baker)

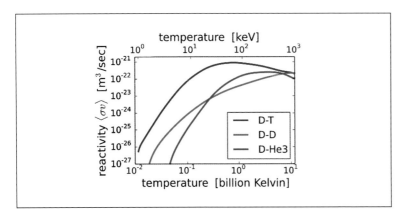

ABOVE Relative temperature and energy levels for fusion reactions based on deuterium-deuterium, deuterium-tritium and deuterium-helium-3 chain.

(Via David Baker)

naturally occurring isotope and has the longest half-life of all, at 4.468 billion years. But this isotope requires neutrons of at least 1MeV to cause fission. While most neutrons produced in fission have higher energies, they rapidly lose energy in collisions, which quickly brings them below the threshold for fission so that sustained reaction is impossible. However, some fission does occur and the energy released can contribute appreciably to the total energy produced in nuclear weapons.

Just 0.7% of all uranium is of isotope 235 and this and plutonium-239 were the only nucleons used in early nuclear weapons. Uranium is weakly radioactive, releasing alpha particles, and is chemically toxic, producing serious damage to the human body, particularly the kidneys and the brain, and is known to increase the risk of cancer. Nevertheless, we live with uranium as an important part of the Earth's structure and it is arguably the mechanism by which the Earth has produced advanced life forms.

With an average concentration of about three parts per million, uranium is 40 times more abundant than silver in the Earth's crust, with total deposits estimated at about 100,000 billion tonnes, of which 53,000 tonnes are mined each year. More than a quarter of this total comes from Australia. It is the radioactive decay of uranium (along with thorium and potassium) that is responsible for the Earth's heat, and for moving the continents around in the process known as plate tectonics. This process has

consistently remoulded the surface of the Earth, creating niches that scientists now believe are vital for encouraging the development, and proliferation, of evolving life forms.

Naturally occurring nuclear reactions are known to have happened in the Gabon region of Africa 1.8 billion years ago, when the average isotopic abundance of ^{235}U was 3% and not 0.7%. This is the level necessary for nuclear reactions to take place subject to the right conditions, essential for which would have been some form of cooling to regulate the 'reactor' into producing thermal energy rather than an explosion. Studies of the geological remains of the strata today reveal the former presence of 16 reactor zones which worked for a period of 150,000 years on the basis of 30 minutes on, 2 hour 30 minutes off, regulated by the water-saturated sandstone bed. Calculations show that these natural reactors would have produced an average 100kW due to the compression caused by continental motion during the Proterozoic.

To produce nuclear weapons uranium-235 needs to be enriched so that the natural uranium is 90% U-235 rather than the 0.7% found naturally. Enriched uranium is 70% denser than lead and exceptionally hard, but the enrichment process itself produces a waste product with reduced U-235 content of less than 0.3%, and this is the depleted component used in munitions for armour-piercing shells and bullets. Because it is pyrophoric it ignites spontaneously on contact with the air and pyrolises the interior of tanks and armoured vehicles penetrated by these munitions.

Low enriched uranium has about 20% concentration of U-235 but highly enriched, or 'weapons-grade', uranium has at least 85% concentration and some concentrates have been prepared with 97% concentration. The uranium bomb used against Hiroshima had 84kg (185lb) of 80% enriched uranium but, theoretically, a similar bomb would require only 50kg (110lb) of 85% enrichment for a sphere 17cm (6.7in) in diameter.

The earliest forms of enrichment were electromagnetic isotope separation (EMIS), whereby large magnets were used to separate ions of the two isotopes. Quite rapidly, the two favoured methods of enrichment became gaseous diffusion and gas centrifuge separation. In gaseous diffusion uranium hexafluoride is passed through semi-permeable membranes to produce a slight separation between molecules of U-235 and U-238. The lighter molecules containing U-235 penetrate the barrier slightly more rapidly, and with enough stages significant separation can be accomplished. Both EMIS and gaseous diffusion require huge amounts of electricity.

While high heat consumption is essential, the thermal-diffusion process is very simple and a production plant containing 2,100 columns, each about 15m (49ft) long, in operation for less than a year provided less than 1% ^{235}U. Each column comprised three tubes with cooling water circulated between the outer and middle tubes, and the space between the inner and middle tubes filled with uranium hexafluoride (UF_6).

Gaseous diffusion was highly developed and used at the Oak Ridge facility to produce both high-energy and low-energy uranium, but since the 1960s US facilities have been used primarily to produce commercial low-energy uranium, and the last remaining high-energy uranium facility was shut down in 1992. China and France retained their diffusion plants but those in the UK were closed and dismantled. Russia converted from diffusion to the more efficient centrifuges.

In the process involving gas centrifuges, a large number of rotating cylinders provide separation through centripetal force, whereby the heavier U-238 molecules move tangentially toward the inside wall of the cylinder, leaving the lighter U-235 in the centre. This is a much more efficient process and overtook the preference for gas diffusion as the technology and the operating methods improved. It uses much less electrical power, despite the multitude of stages needed to obtain the degree of enrichment required for weapons-grade uranium.

In the general process of producing weapons-grade uranium, a mass of 8,050kg (17,750lb) of natural ore with 99.3% U-238 and 0.7% U-235 will, after passing through the complete enrichment process, produce 2,205kg of enriched uranium consisting of 96.4% U-238 and 3.6% U-235, plus 7,050kg (15,545lb) of depleted uranium waste.

Another source

Plutonium had been discovered as recently as 1940 at the University of California, Berkeley, by a team led by the renowned chemist Glenn Seaborg. Uranium 238 was bombarded with deuterons, heavy hydrogen nuclei containing a neutron as well as a proton, which first produced neptunium, atomic number 93. This decayed using beta rays, which raised the atomic number to 94 and formed a new element. The convention had been to name the elements by planets and so, after uranium (for Uranus), Seaborg chose the next, and most recently discovered planet, Pluto.

BELOW Albert Einstein (1879–1955) made profound connections between matter and energy which formed the basis for the physics of nuclear energy and of bomb-making technologies, extending the transmutation of matter from the molecular to the atomic level. *(F. Schmutzer)*

RIGHT A powerful and articulate advocate of nuclear weapons, Edward Teller (1908–2003) would influence the decision to build thermonuclear bombs and, paradoxically, to persuade President Ronald Reagan to begin the space-based anti-ballistic missile programme known popularly as 'Star Wars'.
(LLNL)

There are 19 known isotopes of plutonium, with the longest-lived being Pu-244, but Pu-239 is the most fissionable.

Unlike uranium, which has to be enriched to weapons-grade quality, plutonium is produced in reactors by bombarding ^{238}U with neutrons from the chain reaction. Because each fission produces only two neutrons on average, the management of the process is crucial. The amount of neutrons necessary to irradiate useful quantities of U-238 are managed through highly efficient instrumentation and precise calculation. A typical production reactor produces about 0.8 atoms of plutonium for each nucleus of ^{235}U that fissions. Typically, one 1gm of plutonium is produced for each megawatt-day of reactor activity, and light-water reactors make fewer plutonium nuclei per uranium fission than graphite-moderated production reactors.

The plutonium has to be extracted chemically in a reprocessing plant, calling for a complicated process that requires the handling of radioactive materials. Because of this it is handled by robots, or by human operators using remote manipulating equipment. Because it uses hot acids to dissolve short-lived radioactive fission products it is inherently dangerous and calls for particularly robust methods of disposing of the

waste. This usually requires very large quantities of concrete shielding and any processing plant will inevitably vent some radioactive gases to the atmosphere.

Fission itself is possible in a number of ways but there is only one that is practical for the release of nuclear energy. The very process itself is defined by Albert Einstein's most famous equation $E=mc^2$, expressing the interchangeability of mass and energy times the square of the velocity of light, and stipulating that wherever there is a sudden change on the level of the nucleic mass the liberation of energy will result. The very reason why nuclear fission is possible is that in the process itself there is a liberation of neutrons to sustain a reaction.

The fission process generates three products: lighter nuclei known as fission fragments, neutrons referred to as fission neutrons, and energy. Taking U-235 as an example, the fission cycle can be expressed as: U-235 + n → fission fragment + 2–3 neutrons + energy. A total of 60 or more isotopes (fission fragments) are formed during the fission process, corresponding to the 30 or so different ways in which the heavy nuclei can split into two lighter nuclei. Most of these fragments are radioactive, emitting beta and gamma radiation. Each fission product goes through three stages of beta decay before becoming stable, which collectively will produce about 200 different isotopic species, most of which are radioactive.

The number of neutrons produced in the fission process varies with the different ways in which the nuclei are split. The energy of fission neutrons ranges from quite small values up to 14MeV or more, the majority having energies of 1–2MeV. The average number of fission neutrons produced varies too according to the initiator used, varying from 2.51/fission for U-235, to 2.60 for U-233 to 2.96 for Pu-239.

The emitted neutrons can be divided into prompt neutrons and delayed neutrons, the former being released in approximately 10^{-12} seconds of the initiating process while the latter continues for several minutes. For U-235, the prompt neutrons constitute 99.25% of the total fission neutron population, and 99.75% for Pu-239. Obviously, because of the time-scale involved delayed neutrons play no part in the reactions. However, they are highly significant

in nuclear reactors, where energy release is controlled and the fission rate is relatively slow.

Approximately 200MeV of energy is produced in each act of fission and this holds good for U-235 and Pu-239. Based on these figures, the energy theoretically available in the complete fission of 1kg (2.2lb) of U-235 (2.6×10^{24} nuclei) is 8.4×10^{20} ergs (where 1 erg equals the amount of work done by a force of one dyne exerted over a distance of 1cm/2.54in; one dyne equals 10 micronewtons). However, only about 75–85% of the energy so liberated is available to produce blast in a nuclear explosion as some of that will escape in the form of thermal energy and gamma rays.

Because of this, the theoretical energy from 1kg (2.2lb) of U-235 equating to the liberation of 20,000 tonnes of TNT equivalent (20KT) would produce a blast effect of about 17KT, or 19KT for Pu-239. The energy yield per fission is equivalent to about 6.6×10^{-24}KT for uranium-235 and 7.3×10^{-24}KT for plutonium-239.

The maintenance of a fission reaction is ensured only when another neutron capable of causing fission is produced. But not all the fission neutrons are available to sustain a chain reaction. In a weapon system, an important source of neutron loss is leakage or escape from the reacting system, so that many neutrons escape before being captured by the fissile nucleus. Some are lost by parasitic capture, that is by non-fission reactions of various kinds, either by the fissile material itself or by other nuclei.

The fraction of neutrons escaping can be decreased or increased by increasing the mass. Because neutrons are produced, by fission, throughout the whole system but only lost from the exterior surface, the probability of escape will decrease as the volume/area ratio of the system is increased, and for a given geometry this can be achieved by increasing the physical dimensions. The critical mass for a chain reaction is dependent on the nature of the fissile material, its shape and several other factors.

If v is the average number of neutrons produced in each act of fission and l is the average number of neutrons lost by escape or other means, then v-l is the number available to sustain fission, which in the following can be represented by k, so that $k = v$-l. For every neutron causing fission in one generation, k neutrons will be available to cause fission in the next generation. Hence, k will be less than unity in a subcritical system but will be unity (k = 1) in a critical mass and greater than unity in a supercritical system.

ABOVE Scientists and staff at the Clarendon Laboratory, University of Oxford, in 1936 when so much theoretical work was being conducted in the UK that would lead to support for the Americans to build a bomb. The contribution made by Clarendon is frequently missing from popular accounts of this work. *(University of Oxford)*

Chapter Two

The American bomb

Forged in war and built to deter aggressive forces in Europe and the Far East, the United States mobilised a vast undertaking that would provide the world with the first nuclear weapons – and the irreversible journey to an Atomic Age began.

OPPOSITE A cleaned-up image of the fireball as it appeared to observers witnessing the world's first nuclear explosion. President Harry Truman had specifically asked for it to be no later than this day, when the Potsdam conference to settle post-war European boundaries took place. *(Jack Aeby)*

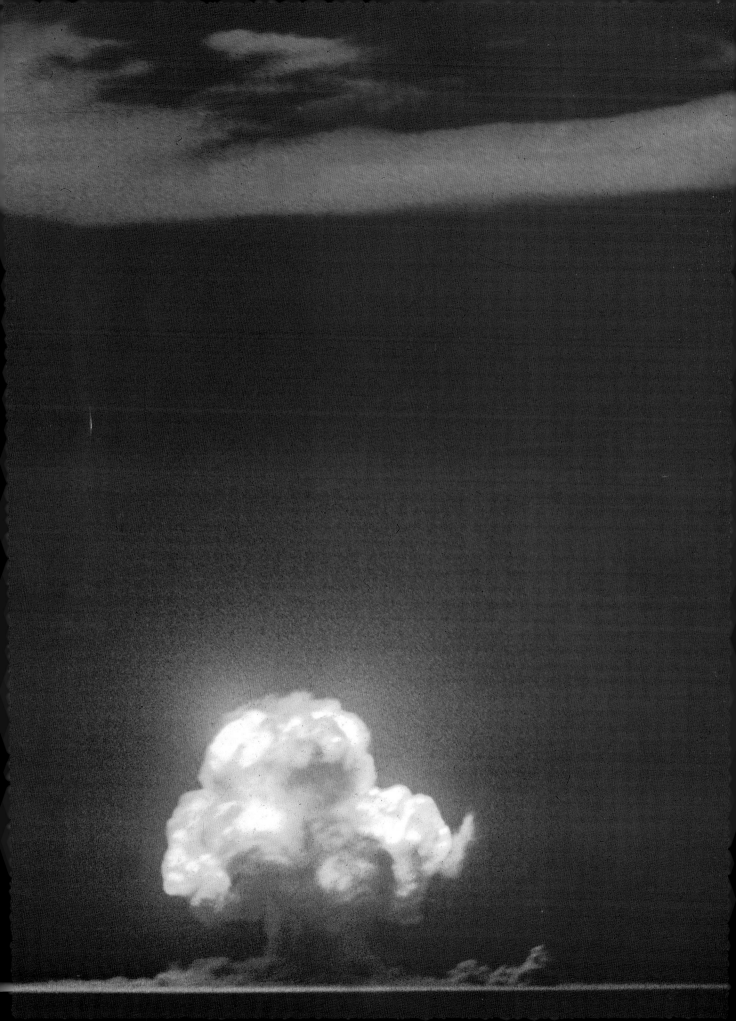

The road to nuclear fission, for peace or for war, began when Rutherford split the atom and when he discovered the proton, and when Maxwell discovered the neutron and measured the relative mass and charge of all three particles. But almost immediately physicists were attracted to the idea of liberating this energy in a nuclear reaction. The Italian physicist Enrico Fermi tried to create new elements and bombarded uranium with neutrons in an experiment that took place in Rome in 1934. He even went as far as to name these [93]Ausonium and [94]Hesperium but in fact, as suspected by several of his contemporaries at the time, he had inadvertently fissioned uranium and measured the fragments as

separate elements, although it was several years before that was realised.

Shortly after the Fermi experiments Otto Hahn, Lise Meitner and Fritz Strassman started work in Berlin on experiments to prove nuclear fission was not only feasible but could be achieved. Meitner was a Jew and fled to Sweden in 1938, as had her physicist nephew Otto Frisch, but the two maintained a correspondence with Hahn in Germany and learned that he had carried out the first demonstration of induced nuclear fission in history.

He bombarded uranium with neutrons, producing barium. Disbelief and doubt pervaded until the calculations could be made. Barium has only 40% of the mass of uranium and there were no accountable methods of radioactive decay that could account for the loss observed. The critical experiment took place on 16–17 December 1938, a month after Hahn had visited Copenhagen to see his old mentor Niels Bohr, along with Lise Meitner and Otto Frisch.

Published in the German journal *Naturewissenschaften* on 6 January 1939, the news electrified the world of physics and Bohr crossed the Atlantic to lecture at Princeton before moving on to Columbia to meet with Fermi and his team, advising them to look for energy released in nuclear fission of uranium nuclei from neutron bombardment. Conducted in the basement of Pupin Hall, a Columbia University team performed the first nuclear fission in the United States on 25 January 1939.

From this came the idea to Hungarian physicist Léo Szilárd that neutron-driven fissioning could be used to create a chain reaction, resurrecting an idea he had had in 1933 but which he had been unable to demonstrate with light atoms rich in neutrons. The theory seemed immutable, that if the number of secondary neutrons produced in a chain reaction could be greater than one an almost unlimited quantity of energy could be released – for endless power or for a force of such devastation that nothing like it had ever been imagined as a practical reality.

The concept of a chain reaction owed much to chemistry, where they were known to exist, but the relevance of a nuclear chain reaction was all the more emphatic given the prodigious amounts of energy that would be released. To investigate this possibility, Szilárd and Fermi spent most of the summer working out the assembly of a nuclear pile, or reactor, to control such a process. Natural uranium would be used as the fuel so that they would benefit from the lower-energy neutrons that would be more likely to 'find' the apparently larger atoms due to quantum theory.

Largely attributable to the work of Max Planck, the refined quantum theory used by Szilárd and Fermi had arisen in the early part of the 20th century and been significantly modified over the following decade. In essence, and very much simplified in description (for words are an inadequate stumbling block compared to the sheer beauty of the mathematics), quantum mechanics explains the state of existence of a system according to a set of wave functions known as state vectors in a matrix of complex vector space. It allows only the probabilities of consequences from particular experiments and, as such, the absolute measurement defies the observer. It allows only an approximation on a set of probabilities, the outcome of which will be the same as a parallel process defined through measurement but for which there can be no certainty in an absolute or consequential sense. The probabilities thus obtained are a product of experiments.

This difficulty is compounded by the interaction between the wave function of the particle observed and the immediate interaction with the instrument being used for the experiment, which changes the very wave form of the diagnosed particle, compromised by the diagnostic tool. In which case the absolute state can never be determined because the probability is based on a consequence, rather than a calculation, and the 'absolute' answer sought could only be found if there were no diagnostic tool – which is impossible within a frame in which time exists. Which is why quantum mechanics explains the timelessness of particle existence and does not quote definitive values but merely assigns a probability distribution.

Despite several decades of research into attempting to find a more deterministic explanation of quantum mechanics by aligning it with the classical concept of causality, it was largely due to the work of Niels Bohr and Werner Heisenberg that it is generally accepted today that such a philosophical destination for the solution is a myth. Albert Einstein too was reluctant to accept the causality principle but the embodiment of quantum theory into the work of Fermi and Szilárd was essential in the striven search for chain reaction. They, as the majority of others since, recognised that relativistic quantum mechanics can be used to describe most chemical reactions and that it was essential to untangling the reaction probabilities in a reactor.

RIGHT Danish by birth, Niels Bohr (1885–1962) made many contributions to physics. In the Second World War he was smuggled out of his occupied country to Sweden, from where he was spirited away in the bomb bay of an RAF De Havilland Mosquito. Crammed in, he failed to put on his headgear with its earphones and did not hear the pilot tell him when to use his oxygen mask as the aircraft reached high altitude. Bohr passed out and only came to when the aircraft descended to land in England. *(Via David Baker)*

Fermi decided to use graphite as a moderator with which the high-energy secondary neutrons would collide, slowing them down and allowing a 'throttle' to be applied to the magnitude of the reactions. In this way he hoped to demonstrate

BELOW Niels Bohr (left) next to James Franck – a German physicist who did much to unravel the laws by which the electron impacts the atom – with Albert Einstein and noted American physicist Isidor Isaac Rabi. *(Via David Baker)*

a sustained production of thermal energy in a controllable way that would also create new fission products. But the realisation that this technique could be used to make a fission bomb troubled Szilárd and several others, who were concerned about the way war clouds gathering in Europe might encourage Germany to do just that and utilise such a device to threaten or subjugate other countries.

A small group gathered to discuss this likelihood and determined that it was very real. Accompanied by fellow physicist Eugene Wigner, Szilárd drove out to Long Island, New York, to visit Einstein on 12 July 1939 and the former dictated a letter addressed to the Belgian Ambassador – as a means of communicating it to President Roosevelt faster than they could achieve themselves – which was duly signed by Einstein as well. Hitler invaded Poland on 1 September 1939, and on 11 October Roosevelt received the letter from Einstein and Szilárd urging priority development of an American bomb so as to deter Germany from developing or using an atomic bomb of their own. When approached about the possibility of such a bomb, Einstein had famously declared, 'I did not even think about that.'

Urgency realised

While reassuring Americans that the United States had no intention of going to war with Germany, Roosevelt was sufficiently concerned about the potential consequences to set up an Advisory Committee on Uranium, chaired by Lyman James Briggs and including Szilárd and Edward Teller. Several were sceptical about the possibility of this committee producing a bomb, but $6,000 was found to support Szilárd's work. The committee was replaced by the National Defense Research Committee in 1940, followed by the Office of Scientific Research and Development in 1941, but the pressure supporting the scientists in carrying out work on reactors and the theoretical possibility of a bomb came from the United Kingdom.

During March 1940, Otto Frisch and Rudolf Peierls prepared a memorandum while working at the University of Birmingham, England, providing calculations for the critical mass required to make an atomic bomb. They postulated a sphere of U-235 and determined that with as little as 1kg (2.2lb) a chain reaction could be induced which would instantly release energy equivalent to several hundred tons of TNT (trinitrotoluene). The memorandum was delivered to Mark Oliphant, who since 1937 had been Poynting Professor of Physics at Birmingham, who passed it along to Sir Henry Tizard, chairman of the Committee for the Scientific Survey of Air Warfare (SSAW).

A committee named after its chairman George Thomson was quickly renamed the MAUD Committee at a perilous time in Britain's affairs: Germany had attacked the Low Countries and was sweeping through France, and it seemed as though Britain would soon be overwhelmed by the victorious German Army. The name MAUD was derived from a telegram sent by Niels Bohr in Denmark to Otto Frisch on the German invasion of Denmark, asking him to pass a message to his housekeeper Maud Ray and to physicist John Cockcroft.

Cockcroft had worked with Ernest Walton under Rutherford, after the latter had conducted experiments into disintegrating nitrogen atoms with alpha articles released from decaying radium atoms. Cockcroft and Walton went on to conduct the first disintegration of an atom using a lithium target to produce alpha particles on 14 April 1932. Security was so tight around the MAUD Committee that at first Frisch and Peierls were barred from participating, but their previous contribution had been so important that they were allowed to carry out work on the separation of uranium.

Listing only eight people including Thomson, the MAUD Committee developed a plan to lay out the requirements for production of an atomic bomb. The committee was liberated from the SSAW and placed within the Ministry of Aircraft Production during June 1940. Two months later Cockcroft, now as assistant director of scientific research in the Ministry of Supply, led a mission to the United States to trade technical and scientific secrets in return for its assistance in the war against Nazi Germany.

The mission remained in the United States for four months and provided the Americans with every major technological and scientific invention which had not yet entered into the

everyday activity of any other nation, including the jet engine, radar, all the work carried out on the potential for an atomic bomb, the proximity fuse, superchargers, special gunsights, rockets, unique submarine detection devices and – probably the most prized of all – the cavity magnetron, which US historian J. Phinney Baxter III noted 'was the most valuable cargo ever brought to our shores'.

The work carried across the Atlantic was to prove crucial to America's leap forward with new technologies and to provide the basis on which political decisions in the United States would accelerate development of plans for an atomic bomb. Appointed Prime Minister on 10 May 1940, Winston Churchill made difficult decisions regarding the assignment of limited national resources and opted to encourage the Americans to use their much more substantial assets to develop a weapon, leaving the UK to postpone development of peaceful uses of atomic energy for the post-war world.

The two MAUD reports of 15 July 1941 focused respectively on the development of a uranium bomb and on the use of uranium for nuclear power. It determined that a bomb was feasible and that 12kg (26.46lb) of uranium would create an explosion with the equivalent force of 1,800 tons of TNT. The American physicist Charles Lauritsen carried this information back to Vannevar Bush, director of the Office of Scientific Research & Development

May 17, 1955   E. FERMI ET AL   2,708,656

NEUTRONIC REACTOR

Filed Dec. 19, 1944   27 Sheets—Sheet 25

FIG.3B.

Witnesses:
Herbert Elletonly
Francis W. Turk
Henry K. Johnson

Inventors:
Enrico Fermi
Leo Szilard
By: Robert A. Fernandez
Attorney

(OSRD), which had been formed on 28 June. Faced with a choice between research on an offensive weapon, which might not be ready before the end of the war, and expanded development of a defensive technology such as radar, Churchill chose the latter and gave the bomb to the Americans.

Cockcroft had been shown theoretical work under way in the United States and it was clear that they were some way behind the British in their research toward a bomb. Churchill was keen to get the Americans working on this as a priority, but they were not in the war and had no compelling reason to accelerate their effort, or to realistically support their own scientists. Oliphant flew to the US in August 1941 and discovered that little information had been delivered to the American scientists, so he personally visited the Uranium Committee and persuaded Ernest O. Lawrence to start uranium research at Berkeley Radiation Laboratory. Then he pressed Bush to force the issue with President Roosevelt in a meeting at the White House on 9 October, where Vice-President Henry A. Wallace was also present.

The Americans had an OSRD liaison office in London and pressed the British for even stronger cooperation, but the British were concerned about the absence of US security when in fact it was they who had a mole, who was already sending to the Soviet Union – for the eyes of Josef Stalin – everything that the MAUD Committee delivered to America. That man was Klaus Fuchs, who had been operating as a Russian spy for some time. But cooperation began to reverse after the Japanese attack on Pearl Harbor on 7 December. The Americans seized the initiative during 1942 and began to set in motion a formal programme to develop an atomic bomb.

Under the OSRD, the Uranium Committee became S-1 of that office and all reference to uranium disappeared. By this time the

Germans had mounted the largest land offensive in history by invading Russia on 22 June 1941, and worsening diplomatic ties with Japan indicated to many that conflict in the Pacific Ocean was months away at most. With pressure from the British and the benefit of all their research through the reports of the MAUD Committee, during the meeting at the White House on 9 October 1941 President Roosevelt approved the atom bomb project. The Army would be in charge of the operation, which would use British and American scientists as well as the refugees from Nazi oppression to build the first atomic bomb.

The first meeting of the S-1 committee occurred on 18 December, 11 days after the attack on Pearl Harbor, and heard reports of the three methods being tried for isotope separation of uranium to obtain the required quantity of high-energy U-235. In addition, two separate paths to reactor design had been set in motion and all five lines of development were approved at a further meeting on 23 May 1942. In that month Robert Oppenheimer had been asked to carry out work on fast neutron reactions and in June the Manhattan Engineer District was established by the Army to start moving the programme from the OSRD to full military control and all that this entailed.

Absolute secrecy was vital and security screening of everyone involved in the project was at the highest level. In charge was Colonel

ABOVE LEFT Earlier in the century Max Planck (1858–1947) had done much of the groundwork used by Fermi and Szilárd to calculate the way in which uranium would be used to build the world's first nuclear reactor. *(Via David Baker)*

ABOVE Many famous physicists are seen here, gathered for lunch in Berlin on 11 November 1931 as guests of Max von Laue (1879–1960), the German physicist who discovered the diffraction of X-rays by crystals. From left: Walther Nernst, the chemist who made great strides with solid state physics; Albert Einstein; Max Planck; R.A. Millikan; and Laue. *(Via David Baker)*

BELOW In Britain, a concerted effort to find the optimum way to build a reactor, provide a controlled fission process for power production and perhaps even to make a bomb, had made great strides, involving an international effort. Here (from left) William Penney, Otto Frisch, Rudolf Peierls and John Cockcroft pose for a group picture. *(Via David Baker)*

ABOVE One of the 29 methods of assembling the pile for the first reactor set up in the basement of a baseball field at Stagg Field, Chicago, simply noted as Chicago Pile-1. *(DOE)*

reactors; the only place where plutonium 239 could be produced would be through the nucleic reactions occasioned by stimulated uranium decay through neptunium. As early as April 1941 the National Defense Research Committee set up a project under the control of Arthur Compton from the University of Chicago. By October he had written an encouraging report explaining how nuclear power could be used to propel ships, to produce almost limitless quantities of electricity and to produce a bomb using uranium 235 or plutonium, which had only just been discovered.

Earlier that year, Niels Bohr and John Wheeler had produced a small quantity of plutonium in the 152cm (60in) cyclotron at the University of California, showing that the neutron capture area of plutonium was 1.7 times that of uranium. Something better than a cyclotron was needed to produce the amounts of plutonium required to facilitate further development of a plutonium bomb and in December 1941 Compton was put in charge of producing reactors to convert uranium into plutonium, to devise a means of separating it from the uranium slug and to devise a method of building nuclear reactors. He produced a schedule declaring he would demonstrate controlled chain reaction by January 1943 and a bomb by January 1945.

Throughout 1941 Compton attempted to organise a collegiate group of physicists to put their brains to work on development of a reactor but each wanted it built in their own laboratories at their own universities, and there was little agreement on how it should be designed. Various low-key efforts had been attempted but with little success, so Compton decided to take charge of the operation and to base the work at his own facility at the University of Chicago. Nobody had built a reactor before and there was a degree of uncertainty regarding the effectiveness of Compton's plan, let alone his design, and there were lingering fears of a catastrophic, runaway reaction.

Leslie Groves, a man of little imagination and even less scientific knowledge, but a project manager second to none. Groves had been responsible for the construction of the Pentagon, which would host a separate Department of Defense, the military's replacement for the War Department, when the war was over.

The knowledge that an explosive device through nuclear fission could be made, and that either uranium or plutonium could be used as fuel, stimulated development of nuclear

RIGHT Augustus Knuth poses to display to the camera the adjustment to the wood frame for the piling at Stagg Field. *(DOE)*

Courting reaction

Compton negotiated for the use of an underground rackets court beneath Stagg Field, the university's football ground, which was still used for squash and handball. The

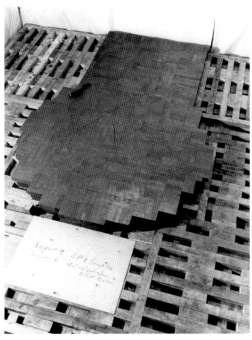

area measured 18m (60ft) by 9.1m (30ft) and was 7.9m (26ft) high. Working with Samuel K. Allison, a physicist of great standing, he brought in all the research teams from Columbia University, Princeton University and the University of California, and, when the Army took over the nuclear weapons programme in June 1942, Compton's reactor work came under the Manhattan Project.

In two months to the day, starting on 15 September 1942, Compton's groups built 16 experimental piles under Stagg Field. Enrico Fermi opted for a spherical pile to maximise k – the neutron multiplication factor – to an anticipated 1.04 that would provide criticality. Preparing for her doctoral thesis, Leona Woods constructed boron trifluoride neutron detectors and assisted with locating a large number of 10cm x 15cm (4in x 6in) timbers acquired from local lumber yards. Graphite of high purity arrived from National Carbide and a few other suppliers and uranium dioxide was delivered from Mallinckrodt in St Louis. A special form of metallic uranium was also supplied.

Meanwhile, the Army had selected a place in the Argonne National Forest, just outside Chicago, for a satellite facility to determine the feasibility of a chain reaction and to work toward development of a bomb. The intention was to achieve critical chain reaction at the underground squash court and move operations

out to the Argonne Laboratory, which was intended to be the first plutonium production reactor plant. As it turned out Oak Ridge would get that honour and operated as X-10.

Back at Chicago University, Fermi was convinced that while prototype testing had demonstrated the feasibility of a working reactor, the squash court itself should be the location of a fully running reactor model without waiting for the Argonne facility. Concerns welled up again and the uncertainties never got any

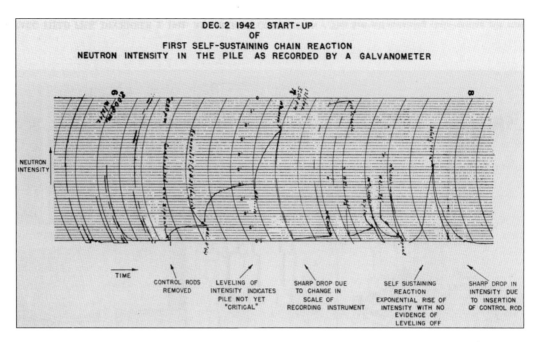

DEC. 2 1942   START-UP
OF
FIRST SELF-SUSTAINING CHAIN REACTION
NEUTRON INTENSITY IN THE PILE AS RECORDED BY A GALVANOMETER

NEUTRON INTENSITY

TIME

CONTROL RODS REMOVED

LEVELING OF INTENSITY INDICATES PILE NOT YET "CRITICAL"

SHARP DROP DUE TO CHANGE IN SCALE OF RECORDING INSTRUMENT

SELF SUSTAINING REACTION EXPONENTIAL RISE OF INTENSITY WITH NO EVIDENCE OF LEVELING OFF

SHARP DROP IN INTENSITY DUE TO INSERTION OF CONTROL ROD

better, with some scientists deeply convinced that there was real danger of irradiating the entire population of one of America's biggest cities with a radioactive fog.

It would take only nanoseconds for released neutrons to start the fission process but Fermi calculated that by using delayed neutrons and by observing the rate at which the fission products were building up it would reach criticality slightly below the level required for a chain reaction. Because there would be spikes in the neutron flux, Fermi believed there would

be time to insert a neutron poisoner and shut it down. But he did not know for sure – nobody had ever done anything like this; it was the world's first working nuclear reactor.

Persuaded by Fermi not to take the project to the university's President and Chancellor Robert Maynard Hutchins, simply because he would not understand the physics involved and would have to deny permission if only to safeguard the university (not to mention Chicago), Compton agreed to the 'experiment'. When told about the plan, James B. Conant, the chairman of the National Defense Research Committee, reputedly 'turned white'.

The entire assembly, known as Chicago Pile-1 (CP1) began on 16 November 1942 with the stacking of graphite blocks inside a cube-shaped balloon 7.6m (25ft) across which was attached to the ceiling on the top surface and flush to three walls on the sides. The remaining side contained a folded curtain area and the whole assembly would contain the powdery graphite dust that covered everything it touched. Approximately 30 former high school students were recruited to the labouring work of assembling what would amount to a structure 6.1m (20ft) high, 1.8m (6ft) wide at the base and 7.6m (25ft) across, containing 5.4 tonnes (6 tons) of uranium metal, 45 tonnes (50 tons) of uranium oxide and 360 tonnes (400 tons) of graphite.

Fermi had expected to need a spherical pile but as the assembly progressed a boron

trifluoride neutron counter measured progress at each level, with readings taken at the end of each 12-hour shift during which two layers were added. As the levels were built, the neutron count began to decrease toward unity, the point of criticality, and it was clear that no more than 57 layers would be necessary. The control rod mechanism worked by inserting cadmium sheets nailed to wooden strips into the uranium blocks, drilled out of the graphite blocks, which caused great clouds of graphite dust to fill the air and cover the floor with the slippery microscopic beads.

The scientists were convinced that by withdrawing the cadmium control rods the fission reaction would go chain and be self-sustaining. On 2 December 1942, at precisely 9:54am, physicist Walter Zinn began to carefully remove the emergency control rod, held by a rope and a pendulum that, upon being severed by the chop of an axe, would swing the rod straight into the pile. Just in case, a technician stood by with a bucket of cadmium nitrate to throw over the stack.

As William Overbeck read out the neutron count on the boron trifluoride detector the sole remaining person on the floor, George Weil, slowly began to withdraw the other rods, leaving only one. The other scientists were waiting, including Leona Marshall the only woman present, and watching in an open gallery above the floor. Fermi ordered the last

rod to be very slowly withdrawn to a maximum 4m (13ft) by a movement of only 15cm (6in) at a time. Suddenly, the over-sensitive automatic cut-off swung the pendulum and slammed the safety rod back in, having been set too low. Time for lunch.

At 2:00pm everybody was back inside the building and the withdrawal of the control rods resumed. At 3:25pm the rod was slowly pushed back in one final time and Fermi did slide-rule calculations, scribbling notes on its ivory-clad back. Breaking out into a smile he declared that the reactor was self-sustaining, albeit producing only 0.5W, enough for a tiny pencil light. But it had been brought to criticality, a k factor of 1.0006, by human hand.

If there ever was one defining moment when the world entered the Atomic Age, this was it. But even at this ebullient moment, celebrated by Fermi producing a bottle of Chianti and sharing it out to plastic cups held in trembling hands, there were dark clouds of concern. As most sidled away, uncertain how they should regard this moment, Szilárd remained, leaning on the balcony rail, gazing down at the pile. Turning to Fermi he shook his hand, held it firm and uttered seven prophetic words: 'Black day in the history of mankind.'

In subsequent work, the power output of CP-1 was raised to 200W although, without any shielding of any kind, raising the output still further risked serious radiation poisoning.

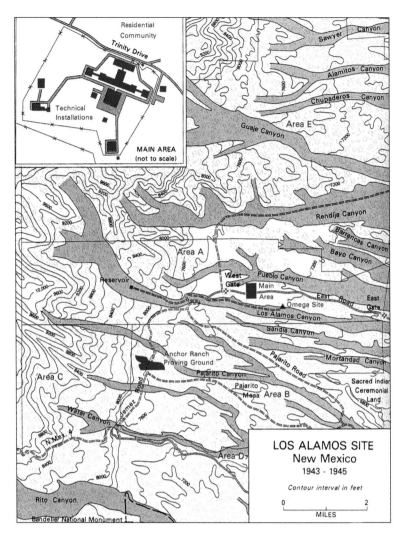

LOS ALAMOS SITE
New Mexico
1943 - 1945

Contour interval in feet

0   2
MILES

LEFT Los Alamos, New Mexico, was selected as the site of the new laboratory to design and build the first fission device, its remote and relatively uninhabited location serving the needs of security and space to expand facilities and test the bomb. *(LANL)*

But a second pile (CP-2) was raised out of the deconstructed materials from CP-1, which had been closed down on 28 February 1943. When CP-2 became operational at the Argonne Laboratory the following month it demonstrated a *k* of 1.055 and was operated continuously without shutting down for the duration of the war. CP-3 began operations on 15 May 1944 and was the first heavy water reactor.

The Atomic Age

While the Americans had been working to extend their commitment, funded activity in the UK fell far behind in a work code named Tube Alloys. Fearing that they would quickly become eclipsed in a field where they had until recently been dominant, on 30 July 1942 Churchill was advised by the minister for Tube Alloys, Sir John Anderson, that it was essential for the British to consolidate their interests in the nuclear field to preserve post-

RIGHT The Clinton Engineering Works at Oak Ridge providing facilities for plutonium research, the Clinton Pile first being designated X-10 in the lower centre. The Y-12 separation plant is upper right, with the K-25 and K-27 gaseous diffusion plants lower left. The facility later changed its name to Clinton Laboratories and was then embraced by Oak Ridge National Laboratories. *(LANL)*

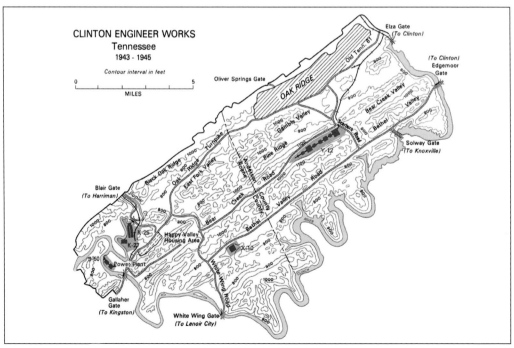

CLINTON ENGINEER WORKS
Tennessee
1943 - 1945

Contour interval in feet

0   5
MILES

war interests. Later that month, Churchill and Roosevelt agreed between themselves to retain that special working relationship but there was nothing in writing.

In March 1943 there was general agreement in the US that the British would be of material benefit to the work and on 19 August Roosevelt and Churchill signed the Quebec Agreement outlining the avenues of cooperation. Under its terms the United States and Britain agreed never to use atomic weapons against each other, never to use them against third parties without the other's consent and never to communicate the secrets of nuclear weapons technology without mutual agreement.

The British involvement accelerated and in December 1943 the government sent Niels Bohr (who had been spirited out of Denmark by British agents), Otto Frisch, Klaus Fuchs, Rudolf Peierls and Ernest Titterton to the US to work on the 'Manhattan Project', as it was now known. In the subsequent, so-called Hyde Parke Agreement of 17–18 September 1944, cooperation was extended to the post-war period until dissolved by mutual consent. More scientists had already followed earlier in the year and while some returned to the UK many remained in the US until the end of the war, by which time the bomb had demonstrated its capacity to change warfare on three occasions.

Meanwhile, the facilities needed to develop the bomb had begun to take shape, primarily at Los Alamos, New Mexico, and at Oak Ridge, Tennessee. Situated on a remote mesa 40km (25 miles) north-west of Santa Fe, Site Y would eventually become the Los Alamos National Laboratory, but in 1943 it was the primary location of all the theoretical and experimental activity which would determine the requirements for industrial development. There had never been a specific plan for volume production of atomic weapons – nobody knew precisely how to manufacture them on a production-line basis – and the explosive potential of the 'gadget', as it was colloquially known, was uncertain.

Site Y, or Project Y as it is referred to in some official documents of the time, involved the acquisition of 22,000 hectares (54,000 acres) of which all but 16% was already owned by the government. But the site required large amounts of water delivered on site together with

electrical power lines brought 40km (25 miles) cross-country. Operations at the site were very firmly under the control of the Army and there was talk of the scientists having to be seconded into uniform. This idea was bound to aggravate existing tensions between the military and the civilian occupants of the site and it disappeared.

Oak Ridge, also known as the Oak Ridge Reservation, and under code as Site X, was chosen as the place where U-235 would be produced through gaseous diffusion carried out at the K-25 site, and electromagnetic separation at the Y-12 plant. The government authorised forced acquisition of 23,000 hectares (56,000 acres) but a further 1,200 hectares (3,000 acres) would be added to the site at a later

ABOVE When Oak Ridge, Tennessee, was selected for the second facility supporting Project Manhattan, it became necessary to clear the population. An eviction notice dated 21 November 1942 was served on a local family demanding their removal from their homes and 77.8 acres of land within ten days. *(LANL)*

ABOVE An exterior view of the graphite reactor at the X-10 site as it appeared in the 1950s after several years of development. *(David Baker)*

returning home from the fields to find a notice nailed to the door or to a tree trunk in the yard. Some families were experiencing their third eviction by federal edict, making way twice before for creation of the Smoky Mountains National Park and for the construction of the Norris Dam. None knew what their eviction was for – they were given only the bland excuse that the federal government needed the land. In July 1943, Groves declared the site a military zone and acrimonious disagreement festered between county officials and the Army, the former having taken out loans to pay for roads and bridges now closed to all but military traffic.

The facility was to be a massive complex capable of accommodating 13,000 people, supported by hospitals, a PX shopping facility, schools, nurseries, medical facilities and recreational buildings. Very little was spared to provide comfort and quality residential homes with all the amenities expected of a modern 1940s lifestyle, which was certainly equivalent to a British lifestyle of the late 1950s! Security was strict, barbed wire fencing surrounded the complex and colour-coded badges directed people to accessible areas according to their clearance and the work they performed.

But initial estimates for accommodation were quickly exceeded and by the peak month of May 1945 some 82,000 people worked at the Clinton Engineer Works, with 75,000 living in the new town specifically created to accommodate the vast majority of the workforce. Oak Ridge morphed into the Oak Ridge National Laboratory, a place where a range of refining activities were conducted, including the construction of a plutonium separation plant known as X-10, which began on 2 February 1943.

Originally known as the Clinton pile, the

date. Known as the Clinton Engineer Works, the site was cleared of 1,000 local resident families, mostly farmers who had settled the area in the 19th century and eked out a living on produce from the land ever since.

The majority were given six weeks to leave, learning of their compulsory purchase on

LEFT The X-10 reactor consisted of a block of graphite 7.3m (24ft) on each side pierced by 1,248 diamond-shaped channels in which rows of cylindrical uranium slugs formed long rods. Cooling air was circulated through channels on each side, and after a period operators would push fresh slugs into the channel at the front face, pushing irradiated slugs out the back into an underwater bucket. *(LANL)*

RIGHT When built, in 270 days and without blueprints, the Oak Ridge K-25 facility would be used for the gaseous diffusion method of enriching uranium as one of three prospective processes, including the Y-12 electromagnetic facility and the S-50 liquid thermal diffusion method. Here, the old moves away for the new. *(LANL)*

X-10 was built about 16km (10 miles) from Oak Ridge and was the world's first true plutonium production reactor when it began operating on 4 November 1943. Built by DuPont, it was designed for a thermal power output of 1,000kW, but the reactor soon exceeded design goals and began operating at 1,800kW from May 1944. Fuelled with natural uranium, the X-10 was air-cooled and moderated by graphite rods. The site had its own research laboratories and support facilities and its staff were housed at the main Oak Ridge town.

Electromagnetic separation of uranium ore was conducted at the Y-12 plant using the technique first developed by Lawrence at the University of California Radiation Laboratory.

The S-1 Committee authorised Stone & Webster to build the facility located at the eastern boundary of the Oak Ridge Reservation, which was set over a sprawling area. The facility enriched U-235 to between 13% and 15% and delivered the first few grams to the Los Alamos Laboratory in March 1944.

LEFT With a floor area of 152,000m^2 (1,640,000ft^2), when completed in 1944 the four-storey K-25 was the largest building in the world. It had a volume of 2,760,000m^3 (97,500,000ft^3) and has been exceeded only by the Vehicle Assembly Building at NASA's Kennedy Space Center, raised just over 20 years later to contain the giant Saturn V Moon rocket. *(LANL)*

RIGHT The principle of gaseous diffusion as used in the K-25 facility. It was based on the basis that molecules of a lighter isotope would pass through a porous barrier more readily than molecules of a heavier isotope. The process required myriads of repetitious cycles, eventually producing increasingly rich U-235 through a series of cascades. *(LANL)*

BELOW One of the K-25 gaseous diffusion cells at the K-25 plant, where 12,000 employees worked. *(LANL)*

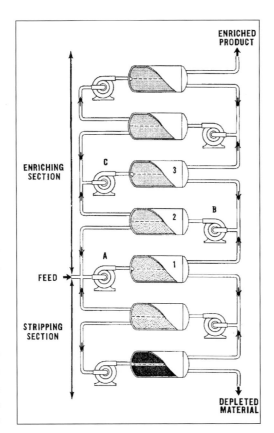

ENRICHED PRODUCT

ENRICHING SECTION

C

3

2

B

A

1

FEED

STRIPPING SECTION

DEPLETED MATERIAL

Gaseous diffusion of enriched uranium was carried out at the K-25 site at Oak Ridge, named after the Kellex Corporation which had been set up as a separate wholly-owned subsidiary to disconnect it from the Manhattan Project, plus the figure '25', derived from the code number for U-235 used universally in the Army during the war. The 600-stage diffusion plant was operated by the Union Carbide Corporation and consisted of a four-storey structure in a U-shape with a total length of 800m (2,625ft), consisting of 54 buildings placed wall-to-wall divided into nine sections of six stages. Each cell could be operated individually or consecutively.

Construction started in October 1943 and the facility was ready for operation on 17 April 1944, but the following year Groves cancelled the upper stages and began work on a 540-stage side feed facility designated K-27. Operations commenced in February 1945 and by April a 1.1% enrichment level had been achieved with the S-50 diffusion plant feeding in. By August, when the last of 2,892 stages

had been put in operation, enrichment levels of 7% were standard, increasing to 23% the following month. The facility incorporated a 235MW coal-fired power station although the majority of the power came through the main grid system.

By early 1946 the plant was producing 3.6kg (7.9lb) of 30% enriched uranium per day with the aim to raise this to 60%. This was achieved on 20 July. The concern that higher refinement could lead to an accident should the uranium go critical and inadvertently result in an uncontrolled chain reaction was met with intense deliberation and analysis before it was decided to proceed with caution.

The plant began producing 94% enriched uranium on 26 November but that process uncovered an inherent flaw in the gaseous diffusion process, in that unneeded U-234 was also enriched as a component of the uranium and made it virtually impossible to reach 95%. Los Alamos had been pressing for the full 95% enrichment but the 94% was good enough for them to work with and the Y-12

LEFT All personnel at the Oak Ridge facility had to take a lie detector test, many employees not knowing the end product of the tasks on which they were working. *(LANL)*

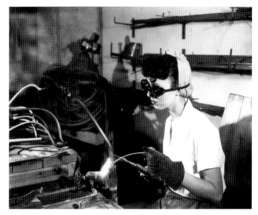

LEFT Most of the activity at Oak Ridge was a big engineering job, the facilities being built serving activities planned by the scientists. Typical of the workforce, this welder was probably unaware of what an atom bomb was. By mid-1942 one-third of the workforce in the aerospace and defence manufacturing industries comprised women. *(LANL)*

LEFT Operators at their calutron control panels at Y-12. Gladys Owens, the woman seated in the foreground, did not know what she had been involved with until seeing this photo during a public tour of the facility 50 years later. Her instructor had told her, 'We can train you how to do what is needed, but cannot tell you what you are doing. I can only tell you that if our enemies beat us to it, God have mercy on us!' *(LANL)*

ABOVE Shift workers clock off and return to camp facilities or nearby towns, sworn to secrecy amid total lockdown. Everywhere notices warned of idle conversation and the workforce believed that security staff were planted to listen out for gossiping workers. *(LANL)*

BELOW A Y-12 calutron-A electromagnetic separation racetrack used to collect U-235 isotopes of uranium. There were nine A tracks feeding eight B tracks with partially enriched uranium, arrangements of giant magnets containing a number of calutrons between them. The calutron sent a stream of charged particles through the magnetic field, deflecting the lighter isotope. With 96 calutrons, each racetrack was 37m (122ft) long, 23.5m (77ft) wide and 4.6m (15ft) high. Some 1,152 calutrons were operated by 22,000 people in a round-the-clock operation. By mid-1945 it was the largest industrial complex in history. *(LANL)*

enrichment activity was slowed right down on 26 December 1946.

The S-50 thermal diffusion plant was set up at Oak Ridge to explore this technique for enrichment, the process having been demonstrated by two German physicists in 1938. It worked on the basis that a mixed gas passed through a temperature gradient will cause the heavier molecules to accumulate at the cold end and the lighter ones at the warmer end, and this was the method behind isotope separation. Some research was continued against suspicion that it would be ineffective, but the Naval Research Laboratory pushed ahead and were encouraged by a favourable report from Oppenheimer to Groves, who approved the design and construction of such a production plant at Oak Ridge. Oppenheimer wanted the output from S-50 to be fed into Y-12.

The design required 2,142 diffusion columns, each 15m (48ft) tall arranged in 21 racks of 102 each. Each column contained three concentric tubes, the innermost nickel pipe carrying steam from the K-25 plant nearby at a pressure of 6,900kPa (1,000lb/in²) and a temperature of 285°C (545°F) flowing downward. The outermost copper pipe carried water at 68°C (155°F) flowing upward. The void between supported the induced separation of uranium hexafluoride gas.

The S-50 plant began partial production in September 1944 and by the following month it had produced 3.8kg (10.5lb) of 0.852% U-235, but technical difficulties, with the pipes frequently springing leaks, delayed full-scale operation. By June 1945, however, it had produced 5,770kg (12,730lb) with all 21 production racks operating. In the cycle of enrichment, S-50 was the first stage, producing uranium to between 0.71% and 0.89% for feeding to the K-25 gaseous diffusion plant, which raised it to 23%, and from there to the Y-12 facility, where it was boosted to 89%.

In early 1943 the US Army Corps of Engineers selected the Hanford site near Richland, Washington, as the location for a plutonium production facility and the construction of a nuclear reactor to produce this element for the Manhattan Project. Three production reactors were built, just north of the Yakima and Snake River junctions with the Columbia River. This vast area occupies an expanse of 1,740km^2 (670 miles^2) and included buffer regions across the river in two other counties.

The site was chosen on the basis that it was remote and isolated from large urban populations in the event of an accident, and it was due to the fear that even Oak Ridge was too close to the conurbation of Knoxville that the Hanford site was chosen, so that it would serve as a back-up to Oak Ridge and as a secondary source of plutonium from uranium. As with other locations, there was an urgent need to clear the area of local inhabitants and

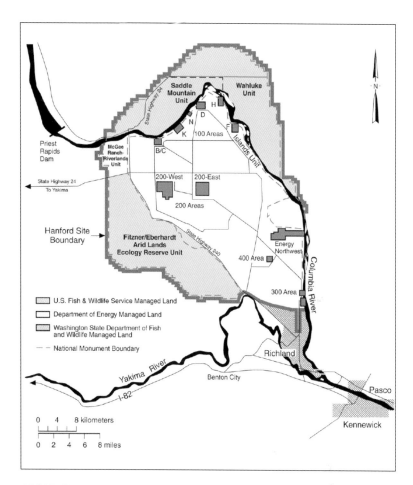

ABOVE The third main complex supporting the A-bomb project was the plutonium facility astride the Columbia River in Washington State. The Hanford site is today dispersed among several agencies and commercial activities and includes a visitor centre. *(LANL)*

LEFT The Richland township where Hanford workers were accommodated. *(LANL)*

ABOVE The world's first large-scale plutonium production facility, the Hanford B reactor during construction in 1944, a massive undertaking which brought in a workforce of more than 50,000 people. *(LANL)*

BELOW When completed B reactor was designed to operate at 250,000kW, rather than the 1,000kW initial output of the X-10 graphite reactor at Oak Ridge. *(LANL)*

1,500 people were relocated, Indian grounds were sequestered and farmers had to abandon their land before most of them could get the crops in, which were ploughed into the ground.

By July 1944 the government had built 1,200 buildings at Hanford and provided facilities for 51,000 people who would work there. It became the third most populous town in

Washington State and duplicated Oak Ridge in the sheer scale of the accommodations and provisions for the workers, including some 900 buses to move them around. The town of Richland became a closed community with tight security and a badging system to allow individuals entry only to areas where they had a legitimate right of access. Paradoxically, efforts were made to make the place less military in appearance and style and the towers and gun posts were less obvious than at Oak Ridge and Los Alamos.

Like so many of the facilities supported by the Manhattan Project, Hanford spawned a wide range of technological innovations and helped with the rapid development of nuclear reactor design, forcing a steep learning curve which gave the country a head start on nuclear power production after the war. Primarily, B reactor contributed as the world's first large-scale plutonium production reactor, designed by DuPont and benefitting from the experimental design details of Enrico Fermi. At first it produced 250MW of thermal energy but that was not its primary purpose, which was to produce plutonium for bombs.

Consisting of a graphite-moderated, water-cooled design, work on B reactor started in August 1943 and incorporated a 1,100 tonne (1,200 ton) graphite cylinder, 8.5m x 11m (28ft x 36ft) in size lying on its side with 2,004 aluminium tubes running horizontally through the centre. Some 180 tonnes (200 tons) of uranium slugs – each 4.13cm (1.6in) in diameter, 20cm (8in) long and sealed in aluminium tubes – went into the tubes. Cooling water was provided at the rate of 110,000 litres (30,000 gallons) per minute, pumped through the tubes and around the slugs.

After overcoming the anticipated process of nuclear poisoning, where short-life fission products capture neutrons, the first plutonium was produced on 6 November 1944 by absorbing a neutron into a U-238 atom forming U-239, which decayed through beta particles into neptunium 239 and then to Pu-239 through a second beta-decay. These plutonium slugs were moved a distance of 16km (10 miles) to chemical processing facilities for separation of the minute quantities of plutonium produced from the remaining uranium and waste products

of the fission process. The initial batch of plutonium, refined through the 221-T plant, was made ready between 26 December 1944 and 2 February 1945 and delivered to Los Alamos on 5 February 1945.

By this date a further two reactors (D and F) had come on line, and within three months regular shipments of plutonium were arriving at Los Alamos every five days, with Hanford providing additional quantities ready for the first live test at Alamogordo and for the first use of the bomb against the city of Nagasaki. Groves had initially planned six reactors at a time when the plutonium was considered for use in the Thin Man bombs, but by mid-1944 the simplified gun-type bomb was found to be impractical for plutonium, which was used in Fat Man where less fuel was required. This reduced the need to three reactors and a reduction in the number of chemical separation plants from four to three.

Time to test

From the physical principles explained thus far it can be said that a critical mass is dependent on the type of material used. Very small quantities of fissile material will not sustain a chain reaction because most neutrons will leak out without splitting other nuclei. The critical factor for a chain reaction is determined by its specific mass and its geometry. The sphere is

chosen because it has the greatest volume for the least surface area, a shape that will minimise to the lowest theoretical value the number of neutrons escaping unproductively. A critical mass of U-235 is 52kg (114.7lb), or for U-233 is 16kg (35.3lb) and for Pu-239 is 10kg (22lb).

There are ways of improving the efficiency, and thereby reducing the critical mass, by covering the surface area of the sphere in a neutron-reflective material which inhibits the escape of neutrons, and appropriate reflectors can reduce this by a factor of up to three. In this way is it possible to reduce the critical mass to

13–25kg (28.7–55lb) for U-235, and to 5–10kg (11–22lb) for U-233 and Pu-239. Reflectors can be of varying thicknesses of natural uranium. It is also possible to lower the critical mass if the density is increased, so that for a fissile material of radius R and a uniform density of p the solution would be: $(p \quad R)_{critical}$ = constant. For a fixed core mass, M, uniformly compressed the density is given by: $p = 3M/4\pi R^3$. As a consequence, the critical mass is proportional to the reciprocal of the square of the density, as in: $M_{critical} = k/p^2$.

The actual design of a fission bomb is determined by several factors, including the type of fissile material and the manner in which it is induced to liberate the maximum amount of energy per unit mass. To achieve that in the shortest possible time, and hence maximise the efficiency, the process must be sustained through a series of stages each identified by the fissioning of nuclei from neutrons in the preceding stage. The stage time is identified as the time between the emission of a fast neutron and its absorption by a fissionable nucleus. The value of a stage (known as a 'shake') is about 0.01 microsecond but the precise value depends on the type of material, the design of the device and the density reached during the explosion.

The energy released takes place over a number of stages, known as generations. A single nucleus would emit 2.5 to 3 neutrons and if 2 go on to produce other fissions the energy released from a device with a yield of 1–100KT would take place after 53–58 generations. About 99.7% of the total energy released would occur in the last 0.7 microsecond. A yield of 1KT will be produced from the fissioning of 1.45×10^{23} nuclei.

Obviously, the yield is proportional to the number of fissioning nuclei, which means that the mass of the assembled material must be several critical masses to sustain a multiplying chain reaction and to avoid the process become subcritical. It can be achieved by assembling several subcritical mass elements or by changing the density and the geometry of the fissile mass. Higher yields can be obtained by increasing the mass above critical and increasing the time the fissile material is held together before the energy is released, stopping the chain reaction.

To prevent premature explosion and the bomb 'fizzling', the fissile material is enclosed within a tamper and can be of the same material as the reflector. Moreover, if the tamper is of the same fissionable material – U-238 – it will contribute to the yield by the fissioning of its nuclei as fast neutrons escape from the interior. All these factors – fissionable material, reflector, tamper, shape and density – contribute to the efficient design of an effective concept, of which there are two: the implosion type and the gun assembly.

In any bomb, two hemispheres of subcritical mass must be brought together symmetrically and uniformly so that a sphere of two hemispheres can become one and compress it into a supercritical mass. In the implosion technique, several separate charges of conventional high explosives are placed uniformly around the surface of the sphere and at equal distances so that when they are detonated the symmetrical direction of explosive energy will create a shock wave passing through the tamper and the reflector to compress the mass to a critical condition.

The conventional charges are known as 'lenses' and are designed to operate with far greater efficiency than any previous explosive since they must focus the blast wave with

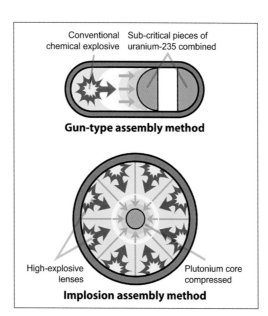

Gun-type assembly method

Implosion assembly method

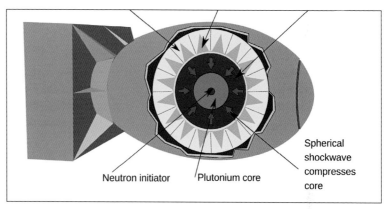

ABOVE The Y-1561 Fat Man plutonium bomb of the type that would be used for the Trinity and Nagasaki tests had a length of 3.3m (10.67ft), a diameter of 1.5m (5ft) and a weight of 4,700kg (10,300lb). The bomb contained 6.19kg (13.6lb) of plutonium, of which about 1kg (2.2lb) fissioned to produce 21KT of energy. The detonation sequence involved 32 explosive columns at the centre of a truncated icosahedron employing a combination of 60% RDX and 40% TNT in the fast explosive and 70% barium nitrate and 30% TNT in the slow explosive. The two were separated by an aluminium sheet that conducted shock waves from the first explosion to the second, nulling out turbulence as it impacted the high density uranium compressing it. *(John Coster-Mullen)*

extraordinary accuracy. It was these strict requirements for a perfectly symmetrical explosion that provided high explosives scientists with some of their most difficult engineering challenges when designing an implosion-type bomb. The inward-directed blast wave increases the density of the fuel by reducing its volume while retaining mass and liquefying the metal before detonation. The

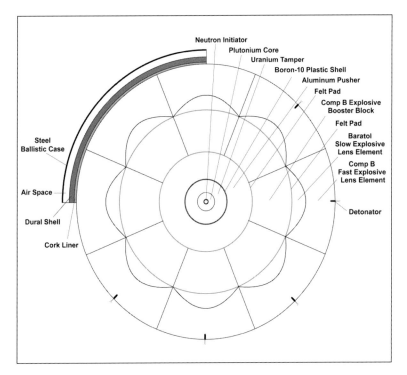

ABOVE A more detailed configuration drawing of the implosion device tested at the Trinity site on 16 July 1945, showing the internal shells that would secure fission in the plutonium core through the neutron initiator.
(Via David Baker)

BELOW The basic components of what was euphemistically referred to as 'the gadget', with the uranium slug encapsulating the plutonium sphere located in its appropriate position late in the assembly process.
(Howard Morland)

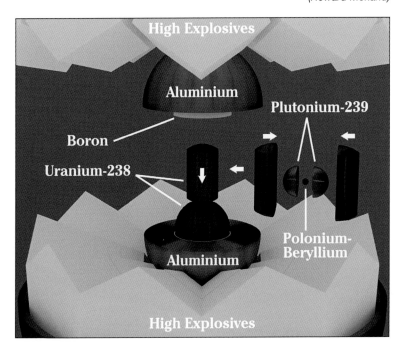

entire assembly inside the encapsulating high explosives is known as the 'pit'.

The gun-assembly type bomb involves at least two and probably more masses of subcritical fissionable material that are brought together by firing a high explosive charge at one end, driving them into a single critical mass suddenly and at great speed. However, reaction occurs because the two (or three) unite, not through compression. This is by far the more efficient system and over time it came to be the preferred design method for some unique applications, one advantage being that it requires a much smaller diameter and is therefore more appropriate for shells designed to fire from tanks or artillery pieces, or for penetration devices. As bomb design evolved after the war, however, most designs favoured the implosion type with either U-235 or Pu-239 materials, largely because the gun-type is almost exclusively appropriate for U-235.

Irrespective of design, the most important part of the fission bomb is the neutron initiator, commonly known as the 'urchin', which will produce an initial burst of low-energy neutrons to begin fission reactions. They must not appear in the system until maximum compression or until the two halves have been brought together. If they appear too soon the bomb will explode prematurely and fizzle out, and if too late the yield of the detonation will be far less than expected. Because the number of neutrons released by successive generations of reaction is proportional to the quantity that began the first generation, the additional neutrons at the beginning will raise the ratio of neutrons at the final generation.

The first initiators were hemispheres of polonium and beryllium located in the centre of the pit. Upon fracturing due to compression the helium nuclei, or alpha particles, from the polonium would bombard the beryllium, producing 30 neutrons per million alpha particles. Polonium has a half-life of 138 days, so the initiators had to be replaced at frequent intervals. By the end of the 1940s actinium was looked at as an alternative but the real solution lay in an external neutron generator in the form of small quantities of tritium or deuterium accelerated by high-voltage generators. When the two materials collide, several tens of millions

A general area map of the Trinity site, located about 100km (60 miles) from Alamogordo, New Mexico. *(LANL)*

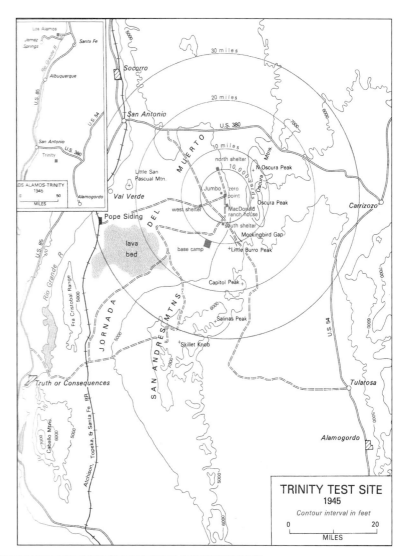

TRINITY TEST SITE
1945
Contour interval in feet

0   20

MILES

of neutrons are emitted and these produce many more reactions at the first generation, which increases the efficiency of the material in the pit.

The story of the development of the first atomic bombs is long and enduring and the countless steps necessary to create the first devices have been told many times. For the first detonation test, the scientists and engineers of the Manhattan Project decided to use a plutonium bomb of the implosion type known as 'the gadget' to personnel working at the pinnacle of activity. There was great debate about the benefits and pitfalls of both the uranium and the plutonium designs and between the implosion and gun assembly concepts. So both would be tried.

The first detonation of an atomic bomb occurred at the Trinity Site on the Alamogordo Bombing and Gunnery Range, New Mexico, at 5:29am on Monday 16 July 1945, a plutonium device delivering a yield of 20KT. Today this is part of the White Sands Missile Range. The Trinity test was a complete success, clearing the way for the use of a U-235 gun-assembly device against the Japanese city of Hiroshima

LEFT After a night of rain and electrical storms, the Trinity test occurred at 05:29:21 local time on 16 July 1945, yielding energy equivalent to that released by 20,000 tonnes of TNT, captured in this dramatic shot of the expanding fireball 16 milliseconds later. *(LANL)*

RIGHT Some time after the detonation, Oppenheimer and Groves pose for a photograph at ground zero, where a new rock, Trinitite, had been formed. (LANL)

mass that was fired down a 7.6cm (3in) tube through the centre of the collection of concentric rings containing 58% of the critical mass of U-235. It fissioned 700gm (25oz) of the total fuel mass of 60kg (132lb) with an efficiency of 1.2% to produce a yield of 15KT. The real significance of the Hiroshima bomb was that it was the first test of a plutonium device, of a gun-assembly design and of an air-drop weapon which could be carried in the bomb bay of an otherwise conventional bomber.

Concern from the Air Force about the effects of the bomb on the aircraft that dropped it stimulated a note from Bethe and Christy describing the effect: 'The sphere immediately heated will reach a temperature of one million degrees Fahrenheit and have a radius of about 30ft (10m). In a very short time of the order of 1/100th of a second this sphere will expand to about 400ft (122m) and cool down to about 15,000°. The ball of fire will rise about 15km (9 miles) in two or three minutes. At the end of this the temperature will have fallen to about 8,000°.

on 6 August 1945. The Trinity test had been in a device situated at the top of a 30m (100ft) tower, but the device used against Hiroshima was the first operational bomb.

Also known as Mk 1 and Little Boy (because of its slim cylindrical shape), the Hiroshima bomb had a plug of uranium of 42% of critical

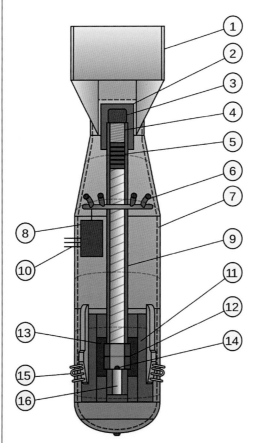

The Y-1852 Little Boy gun-assembly uranium bomb was used for the second test of a nuclear device, on the city of Hiroshima. It had a length of 3m (10ft), a diameter of 71cm (28in) and a weight of 4,400kg (9,700lb). It contained 64kg (141lb) of enriched uranium, most to 89%. Paradoxically, less than 1kg (2.2lb) of uranium fissioned, of which only 0.6gram was converted into energy. Inside, the U-235 was in two parts, defined as the projectile and the target. The projectile was a hollow cylinder consisting of 60% of the total amount of uranium in the form of six rings, at detonation explosively propelled into the target, a slug consisting of the remaining 40%. (John Coster-Mullen)

1 Tail fins
2 Steel gun breech
3 Detonator
4 Conventional cordite explosive
5 U-235 projectile of six rings in steel can
6 Barometric sensing ports
7 Case wall
8 Arming and fusing
9 Barrel 10cm (25in) in diameter, 1.7m (5.5ft) long
10 Arming wires
11 Tungsten-carbide tamper sleeve
12 U-235 target, two rings 38kg (83.8lb)
13 Tamper and reflector assembly
14 Neutron initiator
15 Fusing radar antennas
16 Recess for boron safety plug.

'Because of the small ball of fire and the large distance which the plane can travel before the explosion (7 miles [11km]) the probability of the plane being hit would only be about one in 10,000. The flash of light obtained in the first instance will be as bright as the Sun at a distance of about 100km (62 miles). At the time when it reaches the stratosphere it will still appear as bright as the Moon at 250km (155 miles).'

Bethe and Christy expected the ball of fire to rise to more than 100km (62 miles) but they believed that if it ever came down the radiation would extend over a lateral distance of 100km and 'therefore, be completely harmless'. They did caution that if the device was detonated within 400ft (122m) or less of the surface 'an appreciable fraction of the radioactivity…will make the immediate vicinity of the explosion inaccessible for a considerable time.' As to blast damage to the B-29, that, said Bethe and Christy, would be around '29lb/ft^2' (1.379kPa).

The third bomb, an almost identical copy of the plutonium implosion device used at the Trinity test, was detonated over Nagasaki on 9 August 1945, delivering a yield of 23KT from the fissioning of 1.3kg (2.8lb) of the 6.2kg (13.6lb) of plutonium fuel with an efficiency of 21%, considerably greater than that of the Hiroshima bomb. The success of these three tests, two under operational conditions, was but a prelude to a more robust and determined period of development activity.

The Thermonuclear Age

It is difficult to say precisely when the idea of the hydrogen bomb was first discussed but a logical starting point is on 12 January 1943, when, after lunch, Fermi and Teller were walking back to the building containing their offices. Fermi raised the issue that, now that the atom bomb was under way, it was time to consider a fusion device using lighter elements. Inspired by this notion, Teller went away and did all the necessary calculations on such a weapon, periodically talking it over with Fermi.

The possibility of moving to a hydrogen bomb was raised at the Berkeley Summer Conference that year and was taken up by Oppenheimer, who decided to discuss it on a more formal footing. While Oppenheimer

ABOVE In a rare photograph of the interior components of Little Boy, technicians connect the bomb to test equipment in preparation for the first drop. This was the first nuclear weapon capable of being carried in an aircraft. *(USAF)*

BELOW Little Boy in the bomb pit prior to loading aboard the *Enola Gay*, a B-29 of the 509th Composite Group with modified aircraft and specially trained crew. *(USAF)*

was notionally the head of the conference, Teller did the running during discussions which encouraged support in additional work conducted by Bethe, Urey, Konopinski and Teller himself. Within a few months the prospects were brought up again in Chicago.

Following these informal deliberations, carried out so as not to interfere with development of their primary objective, calculations for the hydrogen bomb were passed through Bethe for critical checking and feasibility. The result

was that the hydrogen bomb was judged to be not only highly feasible but also relatively easy to accomplish. Quickly, the hydrogen bomb developed its own code name – 'Superbomb' – and so it was that the Super became the subject of much discussion as an exciting follow-through once work on the A-bomb peaked.

Oppenheimer, who was looking for something to keep the Los Alamos Laboratory going after work on the A-bomb ended, fastened on the Super as just the kind of project that would do that, pending support from the military and the politicians. But it would have to wait, for just as the Super began to excite interest within the inner core of 'need to know' scientists, work on the physics of the A-bomb revealed the spontaneous fission of Pu-240 which absorbed research into the need for an implosion weapon to liberate the energy in plutonium.

There were now to be two types of A-bomb, based on plutonium and uranium, and attention focused solidly on producing two types of working bomb. Until the successful test of 16 July 1945 further work on a fusion weapon was restricted, but the core group of scientists kept the prospect of an H-bomb firmly in place. And then everything changed. After the dropping of the second A-bomb on Nagasaki the mood of the scientists shifted. With the exception of a very few, most wanted to return to theoretical work, teaching or some other such relatively benign activity.

Most affected by the effect of the bomb on real targets, the morning after the Nagasaki raid Oppenheimer told Teller to 'go home' and

proffered the view that the laboratory should be closed, its work done. Oppenheimer wanted no more bombs, the war was over. It was a terrible but necessary thing they had done and they, and their consciences, would have to live with it. But not Teller, who had seen first-hand the despicable treatment of humans meted out by communists. The spectre of another armed struggle – that between communism and liberal democracy – was to him all too probable. The A-bomb must be perfected, stockpiles raised and work on the Super started immediately.

Four scientists (Oppenheimer, Ernest Lawrence, Arthur Compton and Fermi) wrote a letter to US Secretary of War Henry Stimson urging that work should stop and that the US should ban nuclear weapons altogether. Oppenheimer carried the letter to Washington in person, but before he returned he heard from US Secretary of State James Byrnes that there was no alternative to 'pushing ahead'.

This volte-face from Oppenheimer was all the more indicative of the profound effect Hiroshima and Nagasaki had on him. As recently as January that year, with development of the implosion device well in hand, he had freed Fermi, Frankel, Konopinski, Teller and Horowitz to work on the Super, a group later joined by Landshoff.

But Oppenheimer had had enough. Exactly three months to the day after the Trinity test he left Los Alamos to return to teaching. In an urgent plea for sanity he expressed his deep fear that if the weapon was further developed and stockpiled an atomic war was all too possible, asserting that the people of the world

ABOVE LEFT This image of the explosion was taken from a distance of 10km (6 miles) about two or three minutes after the detonation. *(Hiroshima Peace Museum)*

ABOVE Fires that broke out immediately after the bomb detonated increased over the next several hours, creating a giant pall of smoke slowly drifting away from the doomed city. *(Hiroshima Peace Museum)*

BELOW From the US Bombing Survey report, this map shows the degree of devastation caused by the fires that engulfed Hiroshima on the day the world entered the age of nuclear warfare. *(USBS)*

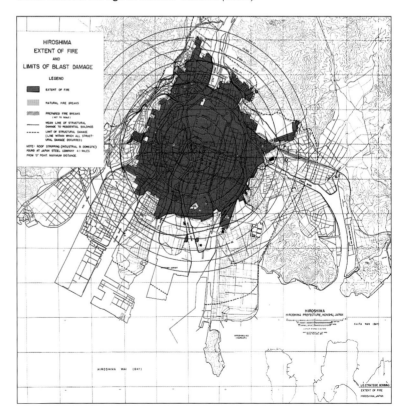

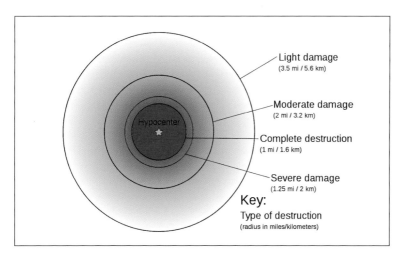

Light damage
(3.5 mi / 5.6 km)

Moderate damage
(2 mi / 3.2 km)

Hypocenter

Complete destruction
(1 mi / 1.6 km)

Severe damage
(1.25 mi / 2 km)

Key:
Type of destruction
(radius in miles/kilometers)

ABOVE The zones of destruction and damage of a Hiroshima-scale bomb showing an area of more than 12.6km² (4.9 miles²) completely destroyed and a further area of 98.5km² (38.5 miles²) from severely to moderately damaged.
(Via David Baker)

'must unite, or they will perish'. Moreover, it was not readily apparent that there was to be any peacetime development of atomic weapons. They had been considered by many scientists to be a necessary, if unsavoury, response to the frightening expansion of Nazi power, which in the minds of many had threatened the very existence of civilised society.

Scientists in general thrive in an open, international setting where free discussion and the willing exchange of ideas and knowledge is paramount to their work. As with so many expanded technological capabilities, the atomic bomb was considered by the majority of those who worked on it as an expedient of its time, and as a 'one-off' to end the war to end all wars – again. Moreover, the older scientists were deeply concerned about the absence of competent young scientists in the United States and yearned to get back into teaching and developing new talent.

One of the most vocal proponents of an expanded development and test programme for atomic bombs was Teller, whose condition for staying required the Los Alamos Laboratory to conduct 12 tests a year and develop the Super as an urgent priority. As the United Nations' Atomic Energy Commission sought every possible way of banning the bomb, the intransigence of the Soviet Union convinced many scientists to return to Los Alamos and resume work on the development of more efficient and bigger bombs – against the day that they would be needed as a deterrent to the development and use of similar weapons by a foreign power.

Deep divisions separated advocates and opponents of the so-called H-bomb. Thermonuclear weapons were so far beyond the explosive yield of the fission devices that their destructive potential was almost beyond comprehension. And that was so because, unlike the limited yield ceiling of fission weapons in the atomic bomb, the hydrogen bomb had no upper limit. It could be made as powerful as desired, beyond the level where it could be used effectively as a weapon and more into the realms of geophysical restructuring of the planet.

The fact that thermonuclear fusion bombs theoretically had no limits worried scientists, who began to see the atom bomb not as a means of ending war but as a tool for military exploitation, a device to usher in a totally new age of conflict where minor powers could wreak unspeakable havoc on the world if they had 'the bomb'. This argument worried the politicians too, but the test of a Soviet atom bomb in August 1949 tipped the scales in favour of a political decision to develop thermonuclear weapons, both as a military expedient and a deterrent and as a political bargaining tool.

RIGHT Stanislaw Ulam (1909–84) was a Polish-American who left Europe in 1939 for a position with the University of Wisconsin-Madison, becoming a US citizen in 1941. Joining the Manhattan Project, it was his calculations that showed it would be impossible for plutonium to work on a gun-assembly bomb; soon he would work with Teller on designing a workable thermonuclear device.
(LANL)

Reacting to the Soviet bomb, President Truman was lobbied by the Joint Chiefs of Staff, claiming that it would be 'irresponsible' not to go ahead with it, and by Secretary of State Dean Acheson, who argued with senior diplomat George Kennan at length in a historic exchange which parallels that of Niels Bohr and Heisenberg. Kennan wanted the United States to restrain its hand and set an example for nuclear non-proliferation, which, argued Kennan, was as equally applicable to non-escalation within the United States as acquisition of atomic weapons by foreign powers.

The Atomic Energy Commission (AEC) supported Kennan's view and in November 1949 voted three to two against development of thermonuclear weapons, which persuaded President Truman to set up a three-man committee consisting of AEC director David Lilienthal, Dean Acheson and Secretary of Defense Louis Johnson. It met only twice and erupted in acrimony and outbursts of polarised viewpoints, but Acheson attempted to mediate the opposing views and presented Truman with a *fait accompli* that Truman had no choice but to go ahead with a thermonuclear bomb.

In essence Acheson was against building the hydrogen bomb, but political expediency and his balanced view of the diplomatic stage left no doubt in his mind that it had to be built, and

that not to do so would be irresponsible, given that the Soviet Union was clearly capable of building its own thermonuclear device and was deemed likely to do so as quickly as possible. When Acheson met with the President he presented the two-to-one decision to go ahead but, there were still doubts that it would start America down a road from which there would be no turning back.

On 31 January 1950 the advisory committee met with Truman, who asked only one question: 'Can the Russians do it?' All three agreed they could, to which Truman responded: 'In that case, we have no choice. We'll go ahead.' It was the first in a series of groundbreaking decisions made after the Second World War which would transform the country and the world. The atomic bomb had been an expedient of war, driven by a concern that Hitler might build a bomb to threaten America. This was different, a decision made to maintain a dominant position in controlling adverse political forces and sending an emphatic message to the world at large, that the United States would not be eclipsed.

The development of the thermonuclear bomb was linked closely with an expansionist policy that sought to maintain pre-eminence for the US military and for political applications on the world stage, and the deterrent policy

LEFT Teller had for long advocated the development of thermonuclear weapons, at one time pressuring for a fusion device rather than the fissioning bombs of the Manhattan Project. Now, with some asking of the magnitude of the fission bombs 'are they big enough' to 'end all wars?', meetings were held for a scientific consensus as to whether they were even possible. (LANL)

RIGHT Along with Teller, Ulam would finally solve the problem of designing the configuration of a thermonuclear fusion bomb by what became known as the Teller-Ulam solution. He developed the idea of using a fission bomb to trigger a fusion reaction – in what were known as primary and secondary stages. But there was a further requirement, to solve the problem of containing the fission reaction before it could disperse through the case. *(Via David Baker)*

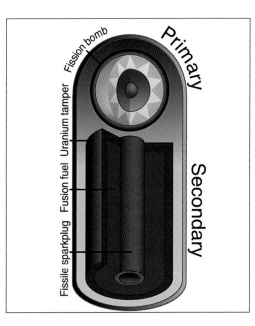

was closely tied in to the decision to build a thermonuclear weapon. This aspiration was embodied in national Security Council Report 68 (NSC-68), which was presented to the President in April 1950. It claimed that a 'new fanatic faith, antithetical to our own' had an objective which was nothing less than 'absolute authority of the rest of the world', and that the Kremlin was intent on 'world domination' and on the 'ultimate elimination of any effective opposition'.

Thus began a major push for a massive escalation in military power and in technological capabilities in a decade which would see the successful development of not only the hydrogen bomb but also miniaturised nuclear weapons, tactical nuclear weapons carried in artillery shells and over-the-shoulder rockets, nuclear mines, ballistic missiles, submarine-launched thermonuclear deployments, weapons capable of hitting targets in Russia within 30 minutes and the beginning of a space programme built around a quantum leap in photo-reconnaissance and intelligence gathering. And the thermonuclear fusion bomb was the trigger that launched all of those possibilities.

The physics of the Super

Nuclear fusion is the opposite of fission: light nuclei are made to fuse by bringing them together at extremely high temperatures of 10–100° million to form heavier elements, a process which liberates great quantities of

energy. There are several different reaction chains possible but the most common are those between deuterium and tritium (D + T), because this reaction proceeds more rapidly at temperatures likely to be obtained by a trigger, that being the detonation of a fission device to start the process.

An alternative fusion chain is between deuterium and deuterium (D + D), but the D + T reaction is 100 times more probable in the temperature range of 1–100KeV (kiloelectron Volts), where 1KeV equals 11.6° million. Moreover, a D + T reaction can be achieved with a lower reaction temperature than for other fuels. The principal reaction being: $D + T \rightarrow He^4$ (3.52MeV) + n (14.07MeV). For comparison, a tritium to tritium fusion provides less energy: $T + T \rightarrow He4 + 2n + 11.4MeV$.

Paradoxically, the energy released per atom is less for fusion than it is for fission. However, because the nuclei are lighter the amount of energy per unit mass is three to four times that obtained from the heavier nuclei in fission bombs. Nevertheless, fission bombs are of relatively low yield and are the simplest form of device. Most of the early weapons were of this type but a combination of fission and fusion reactions came to populate most of the assembled arsenals.

While in this explanation the physical principle of a fusion device appears to have the elegance of simplicity, it was not always so and the controversial path taken to provide a workable solution to the technological hurdles is far from obvious. In the initial calculations proposed by Fermi and so eagerly conducted by Teller, it became clear that the idea of using a fission trigger to detonate a fusion device was unworkable. Early concepts involved surrounding the spherical implosion device with the fusion fuel or placing that fuel at the centre of the fission bomb, in the hope that the thermal energy from the fission release would trigger fusion. It would not, and only through a series of serpentine routes did the now famous Teller-Ulam method emerge.

Born to a wealthy Polish Jewish family, Stanislaw Ulam obtained his PhD at the Lwów Institute but emigrated to America in 1939, where he was invited by Hans Bethe to join the Manhattan Project. It was there that he became

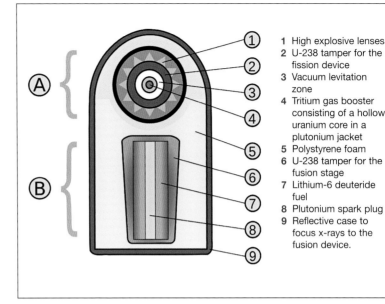

1 High explosive lenses
2 U-238 tamper for the fission device
3 Vacuum levitation zone
4 Tritium gas booster consisting of a hollow uranium core in a plutonium jacket
5 Polystyrene foam
6 U-238 tamper for the fusion stage
7 Lithium-6 deuteride fuel
8 Plutonium spark plug
9 Reflective case to focus x-rays to the fusion device.

aware of Teller's interest in the Super and where he made his outstanding contribution, providing a solution based on a more sophisticated interpretation of the physics involved and of the engineering required to make it a working weapon. The Teller-Ulam solution involved a series of actions in a sequence of separate stages, with the detonation of each leading to the next. The secret was not to use the thermal energy from the fission trigger but to trap the radiation inside for as long as possible so that it would create the next reaction in the sequence.

The solution involved placing a small quantity of deuterium/tritium gas at the core of the primary fission-implosion device which would trigger a small fusion reaction. The abundance of neutrons thus released would induce further fissioning of Pu-239 or U-235 in the primary stage which transfers energy to the secondary fusion stage, and the spark plug which undergoes further fissioning, heating the compressed fuel to start fusion, which adds neutrons to react with the lithium to create tritium for further fusion.

The secondary stage would be wrapped in uranium – enriched or otherwise – or plutonium, and the fast neutrons would trigger reaction in normally dormant U-238, which, under bombardment by these fast neutrons, creates a fission reaction. Further enhancement involves the use of a polystyrene foam as radiation case liner which foams when irradiated with emitted X-rays, pushing against the secondary's tamper, compressing it and starting fission reactions in the spark plug.

One of the peripheral advantages of the fission-fusion-fission cycle is that it maximises energy but with no release of radiation, except

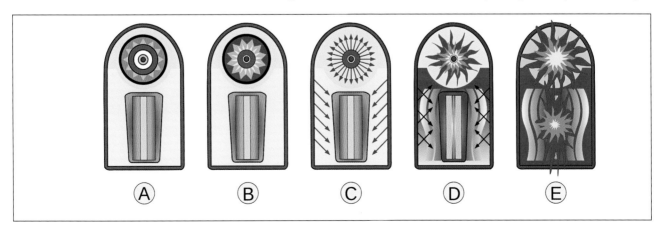

that the fission process within the cycle does release radioactive products and fallout. There are options for adjusting the balance between energy and emitted radiation and if the final fission stage is eliminated and the uranium tamper replaced with lead the explosive force is halved but the fallout is negligible. Conversely, if the tamper is intentionally made thin a large amount of radiation will escape, as with the neutron bomb.

Officially known as an Enhanced Radiation Weapon (ERW), the neutron bomb minimises blast but greatly increases lethal radioactive emissions in the immediate vicinity of the blast. The neutrons that would otherwise be focused inward are allowed to escape, and this has several military applications. In dense urban areas it might be thought beneficial to minimise blast, preserve buildings but intensively irradiate enemy troops, the radiation penetrating armour.

ERW warheads were originally selected for the Lance missile, a tactical weapon deployed in Western Europe for which a conventional warhead was eventually selected. Another application is in anti-ballistic missile warheads where intense bursts of radiation outside the atmosphere would induce partial fissioning in the warheads of incoming re-entry vehicles. ERW warheads were proposed at various times in the 1970s and 1980s when Soviet

deployments surged to challenge existing NATO defensive measures, but the public outcry was so great that they were never deployed.

As technical advancements enhanced the options for bomb improvements, development of fusion weapons would eventually make use of the fusion fuel to enhance fissioning, which would be used as a booster, significantly increasing energy output from what would ostensibly be a standard fission bomb. Degrees of boosting would be used by some countries on the way to demonstrating a full fusion device and would come to be standard design practice as the evolution of bomb design progressed. It was that way with the first fusion device, tested by the United States.

The basic concept of the Teller-Ulam solution was tested during Operation Greenhouse on 8 May 1951, which was a fusion-boosted fission bomb delivering a yield of 225KT, enhanced above the 200KT yield of an unboosted device. Part of a cluster of three tests under Operation Greenhouse, the *George* detonation ignited a small quantity of deuterium and tritium. The following detonation, named *Item*, 16 days later enhanced the yield to 45.5KT from 20–30KT predicted for the basic fissioning device.

The first full-scale test of a thermonuclear device was conducted under the name Operation *Ivy Mike* on 1 November 1952 on

an island in the Eniwetok Atoll. The device weighed 73.8 tonnes and was installed inside a building containing a very large cryostat holding cryogenic liquid deuterium as a secondary stage and incorporating a TX-5 boosted fission bomb with a cylindrical rod of plutonium acting as the spark plug, running down the centre of the cryostat to trigger a fusion reaction.

The entire outer case was a 4.5-tonne uranium tamper, its exterior lined with lead and polyethylene to form a radiation channel for conducting X-rays from the primary to the secondary, starting a fusion reaction in the enveloping deuterium fuel. A steel outer casing 25–30cm (10–12in) thick encapsulated the device, connected to nearby islands in the atoll by a 2.7km (9,000ft) artificial causeway supporting a plywood tube sheathed in aluminium and filled with helium, which allowed gamma and neutron radiation to flow to an unmanned detector station on Bogon. A total of 11,650 people were involved with the test.

Ivy Mike was detonated at 7:15am local time, delivering a yield of 10.4MT, of which 77% came from the fast fissioning of the uranium tamper; it was this that caused large amounts of radioactive fallout. The fireball had a diameter of up to 6.6km (4.2 miles), which was reached within seconds of the detonation. Amid shards of lightning created by the atmospheric effect,

ABOVE LEFT Operation Sandstone was a series of three shots, the first from the Eniwetok facility in the Pacific Ocean. This one (X-ray) conducted on 14 April 1948 was the first and delivered a yield of 37KT. *(AEC)*

ABOVE Development of a production-line fission device was supported with several tests for familiarising troops operating in close proximity to a nuclear detonation, such as this Buster Jangle Dog shot on 1 November 1951. The yield of 21KT was about that of the Nagasaki bomb, the troops being about 9.7km (6 miles) away. *(US Army)*

BELOW Dubbed the 'sausage', the cryogenic equipment for *Ivy Mike*, the world's first thermonuclear detonation, took up a large part of the test equipment, the pipes seen here being for conveying measurement signals. *(AEC)*

the mushroom cloud rose to a maximum altitude of 17km (10.5 miles) in 90 seconds and eventually flattened out at a height of 41km (25.5 miles), spreading to a diameter of 161km (100 miles) on a stem 32km (10 miles) wide.

The crater measured 1.9km (1.18 miles) across and 50m (164ft) deep where the island of Eleugelab had been. The blast wave across the water in the atoll created waves 6m (20ft) high and stripped all the islands clean of their vegetation. Radioactive coral rained down on ships moored 56km (35 miles) away and the entire atoll was radioactive for some time after the event. In the instant of detonation, two new elements unknown before – subsequently given the names einsteinium and fermium – formed in the intense neutron flux.

Ever controversial, Teller disagreed with several key aspects of the test and refused to be in attendance, not wishing to have a possible failure identified with his presence and, by inference, his reputation. Sitting in his office at Berkeley, California, with a seismometer

to hand, he observed its success as seismic waves hit the region. He immediately sent an open telegram to Dr Elizabeth Graves at Los Alamos: 'It's a boy!' *Ivy Mike* was followed on 15 November 1952 by an air-dropped device producing a yield of 400KT, the largest fission bomb ever dropped.

Ivy Mike was a 'breadboard test' – a test in which all the essential elements of a device were designed to evaluate a concept. It was totally impractical as a weapon, but it demonstrated that the scientists and engineers could now move quickly to test a bomb containing a dry lithium-deuteride device, much smaller and closer to the concept for an operational weapon.

The culmination of thermonuclear development was achieved with the Castle series and the *Bravo* shot of 28 February 1954 when the scientists expected a yield of 6MT and recorded an actual detonation of 15MT, the most powerful US detonation ever recorded. It produced a crater with a diameter of 1.8km

BELOW Almost impossible to comprehend in scale, the mushroom cloud from the 10.4MT *Ivy Mike* shot on 1 November 1952, which successfully demonstrated the Teller-Ulam principle of fusion detonation. *(AEC)*

(1.14 miles) with a depth of 73m (240ft). It was followed the following day by the *Romeo* shot, delivering 11MT against an expected 8MT. Of the six Castle tests the fifth also exceeded expectations, delivering 13.5MT versus a calculated 9.5MT.

US tests

The first test of a nuclear weapon took place on 16 July 1945 at Alamogordo, New Mexico, in the United States. Between that date and 23 September 1992 the US conducted 1,032 nuclear weapon tests during which it detonated 1,132 devices. Tests were conducted underground several years before the Partial Test Ban Treaty came into effect in 1963, but some notable shots were included in this total. The largest weapon tested by the US was the 15MT Castle Bravo shot of 1 March 1954 (US time), producing twice the anticipated yield. The Argus I firing in the Operation Argus series on 27 August 1958 was the first nuclear test in space, effected when a 1.7KT device was triggered at an altitude of 170km (110 miles) following launch on an X-17A test missile fired from the deck of the destroyer USS *Norton Sound*. This was followed by two additional high-attitude detonations that year, at 310km (190 miles) and 794km (493 miles) respectively.

Conducted between 25 April and 30 October 1962 and mounted by President Kennedy in response to a series of atomic tests by the Russians – including the 'Tsar Bomba', as it was nicknamed in the West, with a yield of 57MT – the US ran a series of atmospheric shots under Operation Dominic. In this period 31 nuclear test explosions took place, the largest series and the last carried out by the US in the atmosphere, with a combined yield of 38.1MT. Three shots in the series were separately designated Operation Fishbowl, involving the use of Thor ballistic missiles launched from Johnston Island in the Pacific Ocean. The first, on 2 June 1962, failed when the range safety officer had to destroy the missile in flight, but subsequent analysis found that the rocket had been on the correct flight path. The second, launched 17 days later, also failed when the Thor broke apart.

Designated Starfish Prime, the next Thor launched on 9 July achieved its objective with the detonation of a 1.4MT W-49/Mk-4 warhead at a height of 400km (250 miles) during its fall back down toward the atmosphere from a maximum altitude of 1,100km (620 miles). The electromagnetic pulse (EMP) caused by the detonation was far greater than predicted, with damage to electrical installations on Hawaii 1,445km (900 miles) distant. Street lights were knocked out, burglar alarms went off and a microwave link for telecommunications was damaged. A bright aurora lit the sky on the opposite side of the equator and radiation from the explosion persisted in the outer atmosphere for several months, several satellites in low orbits also being damaged or disabled.

The next successful Thor launch occurred on 25 October when a 400KT W50 was detonated at 48km (32 miles), followed by the third and last Thor shot on 1 November 1962, another 400KT test but at an altitude of 97km (60 miles). This was also the last US nuclear detonation in the atmosphere. Several other high-altitude detonations occurred within the Fishbowl series using adapted sounding rockets. Numerous small rockets were launched for each test to sample the atmosphere at various locations close to the detonation and at distant places for effects measurement.

Plans for US nuclear testing began immediately after the dropping of the bombs on Hiroshima and Nagasaki in August 1945,

ABOVE The largest fission detonation of all time took place on 16 November 1952 as the *Ivy King* test, with a yield of 500KT after being air-dropped from a B-36. *(AEC)*

ABOVE Shortly after the first thermonuclear device had been tested, Operation Upshot-Knothole began at the Los Alamos test site, where, on 25 May 1953, a 280mm cannon fired a Mk 9 warhead with a yield of 15KT, demonstrating the rapid diversification of weapons to which nuclear devices could be exported. *(US Army)*

and while the politicians were ruminating over a possible nuclear weapons ban the scientists and the military were determined to proceed with finding out much more about the possibilities laid open by fissile, and eventually fusion, reactions. Initial plans were allowed to proceed while the government contemplated what to do next with this capability.

Operation Crossroads was set up to carry out tests against ships, aircraft and animals and in November 1945 the search began for a suitable test site. In January 1946 the decision was announced to use Bikini Atoll in the Marshall Islands as the test site for the two Crossroads tests, involving 240 ships, 156 aircraft and 42,000 personnel. The first test on 30 June 1946 involved a Fat Man bomb dropped from a B-29 and detonated at 158m (520ft) above a fleet of 90 vessels, with the second on 24 July detonated 27m (90ft) underwater. Both had a yield of 23KT.

Bikini was not acceptable for a sustained test complex since it lacked the dry land area for comprehensive instrumentation and the vast array of equipment essential for monitoring and scientific analysis of each detonation, and it would not be used again before 28 February 1954 when the first hydrogen bomb was detonated, the last of 23 nuclear tests on that

atoll occurring on 22 July 1958. But Bikini had entered the history books, giving its name to the two-piece swimsuit favoured by Romans but shunned by American women until the 1960s as being too indiscreet.

In July 1947 the US announced that it would set up a nuclear weapons proving ground for routine and sustained testing on the Eniwetok Atoll, consisting of 46 islands with a total dry land area of 7.1km^2 (2.75 miles^2) surrounding a 1,005km^2 (388 miles^2) lagoon, 80km (50 miles) in circumference. To clear the site at Eniwetok, President Truman approved the removal of 142 native people to Ujelang, 193km (120 miles) distant, where a village was built to accommodate them. The first tests at Eniwetok were under Operation Sandstone, the first proof tests since the Trinity detonation in July 1945. Second-generation warhead design was evaluated in the three Sandstone tests, with yields of up to 49KT.

Next up were the Operation Greenhouse series, from 7 April to 24 May 1951, four tests with a single maximum detonation yield of 225KT on 8 May as a development test toward the first thermonuclear bomb. This was followed by the now infamous *Ivy Mike* test of 15MT on 1 November 1952. By the time of the last test on 18 August 1958, 43 detonations had taken

LEFT At Eniwetok and Bikini a series of 16 shots took place between May and July 1956 under Operation Redwing, this being the Seminole test on 6 June from the surface of the atoll at Eniwetok, delivering a yield of 13.7KT. *(AEC)*

place on the Eniwetok islands. Ironically, it had been the last test at Bikini (22 July 1958) that contaminated Eniwetok with radioactive fallout. Bikini is too radioactive for people ever to be allowed back but the US government decided to do something about Eniwetok.

Initial evaluation revealed large amounts of caesium-137 and strontium-90 which, with half-lives of 30 years, would be best left in place to decay out. Nevertheless, when the clean-up began in 1978, some 1,000 troops were involved clearing 152,920m³ (200,000yd³) of contaminated soil transported to Runit Island, mixed with cement slurry and pumped into a crater formed by a previous detonation, capped by a massive 45.7cm (18in) thick dome 107m (350ft) across.

But this was only a token gesture. Most of the atoll is uninhabitable, with contaminated sand bulldozed into the lagoon creating higher levels of contamination than would ever be allowed in the United States. The dome is cracked in places and groundwater rises and falls with the tides, washing material out into the lagoon and on to the Pacific Ocean. Already plutonium deposits traceable to Eniwetok are turning up in the South China Sea, 4,500km (2,800 miles) away. A recent survey discovered that contamination is already higher in the

lagoon than in the contents of the dome.

A programme of restoration saw 350 people return to the atoll, consisting of those removed plus their added family members, with around 1,900 now descended from the original native population. The population has stabilised at around 900 people but there is no need for work because none is needed – the food resources come from canned goods supplied by the permanent subsistence aid from the US and so there is nothing to work for. The only real industry on the island is education and the largest building is a church, but it will be several decades before the area is totally clear of surface contaminants, and the threat from rising sea levels and typhoons hammering the dome – possibly to destruction – make it very likely that the entire radioactive waste will end up in the oceans and seas surrounding the Pacific rim.

The majority of nuclear tests have been conducted in the United States at the now famous Nevada Test Site (NTS), where construction began on 1 January 1951, first to prepare a site with an area of 1,760km² (680 miles²), expanded over time to its present area of 3,520km² (1,360 miles²). The site was divided into 28 areas, embracing more than 1,100 buildings, 640km (400 miles) of roadways, 483km (300 miles) of unpaved

tracks, two landing strips and ten heliports.

Located in south-eastern Nevada about 105km (65 miles) from Las Vegas, the areas became a familiar site for off-site residential communities and a tourist trap for bomb-watchers in Las Vegas itself, familiar with watching the mushroom clouds rise over the horizon. The south-eastern corner of the NTS, known as Mercury, hosted the centralised facilities where most of the activity was organised and coordinated.

Tests in the atmosphere were conducted at Frenchman Flat before the 1963 PTBT when it was converted to experimental testing. From that date most tests focused on the Yucca Flat region where underground facilities were provided. Frenchman Flat was the site of 'doom city', a simulated community of mocked-up buildings, food stores, vehicles and mannequins representing the assemblage of a small community under nuclear attack to determine the optimum methods of civilian protection in the event of a nuclear war. Rainier Mesa too was selected for weapons effects testing. In most cases it took up to two years to prepare for a test, operations which cost, in 2017 money,

anywhere from $78 million to $900 million.

The first test at the NTS was under Operation Ranger, the first tests on American soil since Trinity on 16 July 1945. Consisting of five airdrop tests from Boeing B-50 bombers over Frenchman Flat between 27 January and 6 February 1951, they delivered yields from 1KT to 22KT detonated at around 340m (1,115ft). Following a sequence of Greenhouse tests at Eniwetok, activity at the NTS resumed on 22 October 1951 under Operation Buster-Jangle and continued at an accelerating rate. In total, the US conducted 928 nuclear tests at the NTS, of which 828 were underground, all in fact since the last atmospheric test on 4 November 1962.

Routinely during the tests, clouds of contaminated fallout material would wash over the townships across southern Utah, particularly St George where cases of leukaemia, lymphoma and cancers of all kinds, including breast, brain, gastrointestinal and bone, soared far above normal levels. The NTS became the focus for a wide range of protest groups, 15,740 people being arrested over time including Carl Sagan, Kris Kristofferson, Martin Sheen and Robert Blake.

Project Ploughshare

At the urging of some advocates, including Edward Teller, a series of 27 peaceful nuclear detonations were carried out by the Atomic Energy Commission under the Ploughshare programme to evaluate the feasibility of using nuclear explosions to excavate rock and soil for new harbours, to widen rivers and estuaries and to create or widen new or existing canals. Even the Panama Canal was considered for atomic widening!

By 'educating' the public in the benefits to medical science of nuclear physics, and to demonstrate the 'peaceful application of nuclear devices', the AEC sought to 'create a climate of world opinion that is more favourable to weapons development and tests'. The most ambitious proposal was for a series of 22 detonations to cut a road through the Bristol Mountains in the Mojave Desert so as to facilitate the construction of a new highway and a railroad. But it was not all excavation.

On 10 December 1967 an underground nuclear detonation took place at a site in New Mexico 87km (54 miles) from Farmington as an exercise in 'fracking', the extraction of natural gas from rock formations. Adjacent to existing gas wells, the team drilled to a depth of 1,292m (4,240ft) and lowered a 3.96m (13ft) by 46cm (18in) diameter nuclear device that was detonated with a yield of 29KT, almost twice that of the bomb dropped on Hiroshima.

There was great enthusiasm for this method of fracking and engineers calculated that the explosion would provide a cavernous void within the gas-bearing sandstone into which the gas would escape, from where it could be piped to the surface. Today no drilling is permitted within 30m (100ft) of the well, but in 1969 some $8.348m^3$ million ($295ft^3$ million) of gas was extracted, which was found to be contaminated with tritium.

Two further nuclear fracking tests were carried out in Colorado, the first near Rusilon on 10 September 1969, when a 40KT device was detonated at a depth of 2,560m (8,400ft); while considerable quantities of natural gas were liberated, none of it was useable due to contamination. The third test site was at Rio

Blanco, Colorado, on 17 May 1973 when three 33KT detonations took place simultaneously at depths of 1,799m (5,900ft), 6,230m (1,899ft) and 6,690m (2,039ft). None of the extracted oil was useable.

The 27 Ploughshare detonations were carried out between 10 December 1961 and 17 May 1973, with the highest yield being the second such test on 6 July 1962, when an excavation experiment in alluvium liberated a yield of 104KT. The effects testing at these destinations provided military planners with information about the overpressure, the blast pressure and the thermal wave that helped weapons scientists evaluate the consequences of devices of varying yield.

Opposition was high but the scientific pressure to keep on testing for such applications blended into the expanded value for weapons designers. However, the physical consequences were hard to ignore. After the second test at 104KT some 12 million tonnes of soil were displaced in a radioactive cloud ascending 3,660m (12,000ft), ballooning out toward the Mississippi River. Land was blighted, water was contaminated with tritium and communities had to be relocated across a significant area.

BELOW Dubbed the 'runaway bomb', the Castle Bravo shot on 1 March 1954 was the first of America's high-yield weapons which delivered a yield of 15MT, almost three times that predicted. The largest bomb ever detonated by the US, it produced fallout that incurred radiation sickness on residents of nearby islands, on Japanese fishermen outside of the exclusion zone and levels that could be measured far from Bikini Atoll. *(AEC)*

US stockpile

The development of US nuclear weapons stockpiles can be mapped across several separate and distinct periods. Between 1945 and 1950 the bombs were based largely on Fat Man and Little Boy with the heavy bombers of Strategic Air Command the only carriers capable of conveying these weapons for air-drop deployment. Reaction to the announcement on 23 September 1949 that the Russians had exploded their first atomic bomb on 9 August coincided with general expansion of nuclear force capabilities.

In mid-January 1949 the US Air Force had 121 nuclear-capable aircraft, comprising 66 B-29s, 38 B-50s and 17 B-36s. On 1 January 1950 it had 225 nuclear-capable aircraft consisting of 95 B-29s, 96 B-50s and 34 B-36s. As of 1 July 1950 the total had increased to 264. A rapid expansion of the stockpile occurred during the period 1950–55 as the lessons learned from the initial fissile activities fed into production and operational assignments.

Also, the availability of the new, very high yield 'emergency capability' thermonuclear weapons began to emerge, and a new application for tactical warfare now appeared practical due to the miniaturisation of weapons. When considering the tactical and strategic stockpile combined, it is sobering to realise that at its peak in the mid-1960s the United States had more than 30,000 nuclear weapons, all bar a few hundred acquired within the preceding 15 years.

The nuclear age in the United States had forged a completely new industry based around a deterrent which, if used, was capable of resetting the ecosphere of the planet and human life along with it – a far cry from the days when the Manhattan Project had spawned new physics and new technologies, creating bombs which would serve as the base from which all others evolved.

The Mk 1 Little Boy gun-assembly device was first used against Hiroshima and weighed 4,037kg (8,900lb), with a length of 3.2m (10.5ft) and a diameter of 0.71m (2.33ft), employing a 15.2cm (6in) thick case to serve as a tamper and confine the detonation for as long as possible. Five were built. The Mk II Fat Man implosion precursor was a conceptual design capable of carrying either plutonium or uranium fuel but gave way to the Mk II for the Nagasaki mission, of which 20 were assembled. This had a weight of 4,672kg (10,300lb), a length of 3.3m (10.7ft) and a diameter of 1.5m (5ft). With lead acid batteries it could only be maintained at readiness for nine days without changing them and the core radiated so many alpha particles that it had to be removed every ten days, an effort requiring up to 50 men up to 76 hours to complete.

Production of the Mk III began in April 1947 and about 120 were assembled within two years, with the last being retired in 1950. The Mk 4 was based on the Mk III but re-engineered for better dependability, easier handling and improved long-term storage. Begun before the end of the war, development resulted in the first truly operational weapon and 550 were manufactured between March 1949 and May 1951, all retired out of the inventory by May 1953. Some were used on tests.

Its successor, the Mk 5, was produced from 1953 onwards, with retirement complete a decade later. It was designed as a small-diameter implosion device and weight was down to 1,440kg (3,175lb) thanks to a 92-point high-explosive detonation system that produced greater core compression, reducing the weight of the explosives employed. The Mk 5 could be carried by all the bombers in the inventory and some Navy tactical aircraft as well, with 100 fitted to the Regulus and Matador cruise missiles. In all, 240 Mk 5 bombs were assembled.

The first very large-scale production bomb was the Mk 6, which was similar to the Mk 4 but with a lightweight ballistic case and internal sphere. Later versions boasted a 60-point high-explosive system. Several variants were produced with evolutions that would benefit later designs. By August 1952 it was regarded as the standard 1.5m (5ft) diameter strategic weapon, with selectable yields of between 8KT and 160KT. With the same dimensions as the Mk 4 it had a weight of only 3,855kg (8,500lb). Some 1,100 were introduced between July 1951 and early 1955, making it the first mass-produced atomic bomb.

The Mk 7 was named Thor but had nothing to do with the missile of the same name. Production began in August 1952, and 1,820

would be delivered by June 1958. It had a length of 4.6m (15.25ft), a diameter of 0.77m (2.54ft) and weighed 771kg (1,700lb), delivering a selectable range of yields between 0.9KT and 61KT. It had very wide applications, introducing the Air Force to a truly tactical nuclear capability carried on fighters and fighter-bomber aircraft widely deployed at various bases around the world. It was also carried by Honest John and Corporal surface-to-surface missiles and was employed as depth charges with the Navy. It remained the most popular and diverse nuclear weapon in the inventory until retired in 1967.

Inspired by Air Force tests against ships at Bikini Atoll in 1946, the Navy sought a gun-type bomb for penetration strikes. The Mk 8 was the result, a failed concept of which only 40 were assembled with the intention of using it as a cratering weapon carried by the Regulus submarine-launched cruise missile, a role for which it was abandoned. It was the first nuclear device designed for high-g resistance with post-impact detonation.

The Mk 9 was developed for the Army off the back of the Navy's Mk 8, to provide an artillery shell for the 280mm atomic cannon as an interim until a new generation of short-range missiles could be deployed. The cannon could throw the 364kg (803lb) shell a distance of 22.5km (14 miles) and would yield 15KT on detonation. This Hiroshima-size weapon was designed to break up massed troop concentrations. The W9 was produced in 1952–53 and was retired in May 1957. Prompted by development of the artillery shell,

ABOVE With a yield of 9MT, the B53 entered service in 1962 and became the standard high-yield bomb in the Air Force inventory. *(USAF)*

the Mk 9 was a planned successor to the Mk 8 but the project was cancelled in 1955, as was the Mk 11, which lived on as a modified device for Ploughshare.

The great breakthrough in weight that came with the Mk 5 and the Mk 7 in 1952 opened a new era in nuclear weapon deployment that sought to supplement the strategic deterrent with a broad range of attack weapons. They were carried on a wide range of platforms, epitomised by the Mk 12, although only 250 were assembled due to the inefficient use of fissile material. In production from 1954 to 1957 it was retired in 1962 as one of the last fission bombs produced, bomb-makers now transitioning to thermonuclear devices for air-drop and for a new generation of ballistic missiles.

On 28 November 1952 the Los Alamos Laboratory informed the Atomic Energy Commission that a new relatively lightweight thermonuclear device had been designed for use as a missile warhead or as a bomb. The Mk

15 was 3.6m (11.8ft) in length with a diameter of 88cm (34.7in) and a weight of 3,447kg (7,600lb), delivering a yield of 1.69MT. The first American thermonuclear device to enter full-scale production, deliveries began in 1955 and by February 1957 some 1,200 had been supplied. It remained in service until April 1965.

The Mk 17 and almost identical Mk 24 were the heaviest nuclear bombs deployed, weighing up to 19,050kg (42,000lb) with a yield of 10–15MT, the second highest of any bomb developed by the US. It had a length of 7.5m (24.75ft) and diameter of 1.56m (5.1ft), and some 305 were assembled between October 1954 and November 1955, the two types remaining in service until retired in favour of the Mk 36 by November 1956.

A range of different warheads and devices was developed during the late 1950s, and some were produced, but the configuration of the nuclear deterrent and the proliferation of nuclear weapons had outstripped any projection made earlier in that decade. By the time of the Cuba Missile Crisis in October 1962 the United States had a stockpile of 20,000 nuclear weapons, while the Russians had fewer than 7,000. The nuclearisation of US military power would be achieved through the widespread deployment of 5,200 Mk 28 devices. Produced between August 1958 and May 1966, they would remain in service for several decades, finally being retired in 1991.

The B-28/W28 was the first nuclear weapon designed as an entire system with a very wide range of applications, including air-drop by various types of aircraft from heavy bombers to fighters, to cruise missiles such as Mace and Hound Dog, and for a range of selectable yields from 70KT to 1.45MT. Weights varied but all were under 1,052kg (2,320lb) and it was deployed as a multipurpose tactical and strategic thermonuclear bomb and had the second longest production run of any weapon.

Very substantial quantities of other bomb designs were also produced during the 1960s, including 4,500 W31 devices for Honest John battlefield and Nike Hercules surface-to-air missiles, 2,000 W33 nuclear artillery shells for 203mm howitzers and the M115 howitzer, and 3,200 W34 devices for depth charges, nuclear-

BELOW Submarine-launched ballistic missiles evolved at the end of the 1950s as the seaborne leg of the nuclear triad, the Trident emerging as the culmination of several evolutions from the early Polaris missiles. Here, the first Trident C4 is launched on test. *(USN)*

tipped Mk 45 Astor torpedoes and the Mk 105 Hotpoint bomb. With a yield of 9–10MT, 940 Mk 36 bombs were built between April 1956 and June 1958 but all were retired by January 1962.

Produced between September 1960 and June 1962, the colossal three-stage thermonuclear Mk 41 offered a yield of 25MT, the highest ever deployed by the US. It had a length of 3.76m (12.25ft), a diameter of 1.32m (4.25ft) and a deployed weight of 4,840kg (10,670lb). Carried only by the B-47 and the B-52, the 500 Mk 41 devices built came in two versions, a dirty type with tertiary stage encased in U-238 and a clean type with a tertiary cased in lead. The Mk 41 was retired in 1976.

The design of the Mk 41 is interesting if only because of its unique assembly, incorporating a deuterium-tritium boosted primary with a lithium-6 deuteride fuel for fusion in the tertiary stage to trigger a larger third fusion stage, its tertiary stage compressed by the secondary, all encased in a fission jacket.

In use, the Mk 41 would have created a fireball 6.4km (4 miles) in diameter and would have destroyed all concrete structures within 13km (8 miles) at ground zero, flattening residential houses 24km (15 miles) away and producing third degree burns out to a radius of 51km (32 miles). The blast was up to 50% that of Russia's 'Tsar Bomba', which was only detonated once on test, and at a ratio of about 5.2MT/tonne it was the highest yield to weight ratio of any bomb ever built. The Mk 41 had a theoretically limitless yield and it was said that a 35MT device could be developed with a weight of 3,700kg (8,200lb), which could have been carried by the Titan II ICBM.

The Mk 41 was replaced by the Mk 53 and while the Titan II did not get the very high yield version of its predecessor, the Mk 53 (designated W53 for the ICBM role) delivered a yield of 9MT; 60 were produced between December 1962 and December 1963, the last being retired with the missile in 1987 as part of an arms agreement with the USSR. The Mk 53 was also deployed as an air-drop weapon, for which role 340 were built, with the last one decommissioned in September 2011.

The days of large production runs was only just beginning, however, and the Polaris/ Poseidon and Trident SLBM programmes

justified production of 8,780 separate W47, W58, W68 and W88 warheads since June 1960, with the last still in operational use.

Today the US nuclear arsenal is but a small shadow of its once mighty force, with 32,000 weapons at peak in 1967, decreasing by 30% over the following 20 years and by a further 75% to about 5,200 operational warheads by 2009. Every US President since Lyndon Johnson has reduced the stockpile, all because of arms agreements with Russia. In 2017, the United States has 1,740 operational nuclear weapons with a further 4,000 stockpiled but not deployed and 2,800 retired, for a total of 6,800 in the inventory.

The US strategic deterrent is carried by the Trident SLBM force, with 336 Trident II D5 missiles on 14 Ohio class submarines, 450 Minuteman III ICBMs in silos, each missile equipped with up to three 300–500KT warheads, and a strategic bomber force of 20 B-2 Spirit stealth and 76 B-52H aircraft.

BELOW The deployment of tactical and strategic nuclear forces brought nuclear devices to the cruise missile, which experienced a rejuvenation in the 1980s with the Gryphon. Shot into the air from its transporter-erector-launcher by a small rocket motor that burns out when the missile has deployed wings, tail unit and an air intake, a small turbofan engine would propel it to its target at subsonic speed.
(US Army)

Chapter Three

The Russian bomb

⊏━━●━━━━━━━━━━━⊐

Responding to the American bomb, the Soviet Union sought to balance the forces aimed at containing the United States and a race for the biggest bomb and the most powerful nuclear forces fuelled a path that would leave the world breathless in its wake.

OPPOSITE Given the NATO designation SS-15 Scrooge, the RT-20 was developed during the 1960s as the first road-mobile ICBM, with a range of 11,000km (6,800 miles). It would have been a formidable weapon to target, uncoupling the deterrent from fixed silos and moving quickly to evade attack from strike aircraft. However, it was never deployed. *(Novosti)*

Russia's intellectual and scientific interest in atomic physics predates the revolution of 1917 and was continued with enthusiastic vigour during the 1930s, when so much work was being undertaken in Western Europe and in the United States. A closed society to outsiders as well as many of its own citizens, the Russian government trawled the global scientific literature for information about these international developments. But the contribution of Soviet science was to become influential

in directing work toward the possibility of an atomic bomb before it was considered a serious prospect elsewhere.

The story of the Russian atom bomb has been told at length in many widely available publications and it is appropriate here only to relate the highlights insofar as they impinge on the evolution of tactical and strategic systems after the Second World War. Suffice to say that pre-war there were several active research projects under way which conducted fundamental research into nuclear fission, especially after 1939 when several foreign journals described the work going on in the UK, France and the United States. The leverage a bomb would give to any country in possession of atomic weapons was not lost on physicists and on the Soviet government.

By April 1939 independent work in the USSR had verified that each uranium nucleus emitted between two and four neutrons and that this made a fission reaction possible, by using either U-235 or natural uranium with heavy water as a moderator. The Uranium Commission was established in June 1940 and placed under the Soviet Academy of Sciences with work carried out across a broad front. This included searching for uranium resources, the production of heavy water, the building of cyclotrons, laboratory study of isotope separation and the measurement of nuclear constants.

This work, slow and ponderous, came to a temporary halt with the German invasion of Russia on 22 June 1941 but picked up again in early 1942 with news of work on such devices being under way in the UK, America and Germany. Despite the economic pressure of the war, Stalin approved a low level of effort in 1943 under the direction of Igor Vasil'vich Kurchatov. Born in 1903, Kurchatov had been director of the Leningrad Physico-Technical Institute and would be a central figure in the development of both the atom bomb and its thermonuclear successor.

Kurchatov started work on a new proton accelerator at the Leningrad institute in 1933 and within two years he was studying neutron absorption by atomic nuclei, moving on to discover nuclear isomerism. In 1937 he began investigation of resonant neutron absorption by nuclei and drew up plans for work on three

separate, but converging, fronts: chain reaction in an experimental reactor with natural uranium; development of methods of isotope separation; and the early design of atomic weapons employing U-235 and plutonium devices.

By the end of 1943 Kurchatov had 50 scientists working for him, increasing to more than 100 a year later. When President Truman casually mentioned to Stalin at the Potsdam conference on 24 July 1945 that America had a major project under way to develop an atomic bomb, he received a nonchalant response, unaware that the Soviet premier had already ordered an acceleration in Kurchatov's work and that the Russians were already quite close to completing the details of their own atomic bomb.

It is without doubt that the USSR gained measurably from intelligence information passed on by spies, including the nuclear physicist Klaus Fuchs from the Manhattan Project, and that the Russians would continue to gather the results of activity in the UK and the US for some time after the war. As early as 1943 they had obtained a copy of the secret MAUD report. But the originality of basic research in Russia was essential to the rapid pace of progress after Potsdam, when Stalin had expressed his hope that the Americans would use the bomb against the Japanese and bring about an end to the war. Stalin was particularly excited about the prospect of upstaging the Americans by demonstrating that Russia too could build a bomb, and that both Russian and American leaders were already squaring up for the post-war world.

Work on the design of an industrial reactor had begun at the beginning of 1945 and Kurchatov was also involved in the building of an experimental graphite-moderated natural-uranium fuelled pile, known as the Fursov Pile, F1. The first Soviet chain reaction took place in Kurchatov's Laboratory No 2, later the I.V. Kurchatov Institute of Atomic Energy.

Propelled into action by the bombing of Hiroshima and Nagasaki, Stalin demanded a Soviet atom bomb and directed Kurchatov to achieve this by 1948, his henchman Lavrenty Beria being placed in charge of the project. These men were, respectively, the equivalent of Oppenheimer and Groves in America's Manhattan Project, but few of those who played a vital role in the development of the Soviet bomb would achieve the scientific recognition accorded their US counterparts. Meanwhile, Beria decided to develop a remote part of Kazakhstan for the tests and Semipalatinsk was chosen as the site. The main research centre would be located at the town of Sarov in the (then) Gorki Oblast, and renamed Arzamas-16.

The first chain reaction was achieved on 25 December 1946 at F1, a low-power, air-cooled pile that would become the prototype for the Russians' larger, water-cooled, plutonium production plant. F1 initially achieved 100W but subsequent scaling brought it to a maximum intensity of 1kW. The initial pile had been

Chelyabinsk-40, a location that would grow into one of the largest nuclear facilities in the Soviet Union. Situated about 150km (93 miles) from the town of Ekaterinburg, some 72km (45 miles) north-west of Chelyabinsk, the area would become notorious from some of the world's worst nuclear accidents, with large areas contaminated and a great many people affected by the radioactive fallout and explosions which for several decades were denied by Soviet authorities.

The Kazakh Test Site (KTS), near Semipalatinsk in a region near the Aral Sea in western Kazakhstan, was chosen as the site of the first nuclear test (RDS-1), which took place on 29 August 1949, little more than four years after the first American test firing. An implosion device similar to Fat Man with a solid plutonium core, it was mounted to a tower and delivered a yield of 22KT, considerably more than the scientists had calculated. Three days later the Americans detected radioactive debris in the atmosphere during a routine flight with special sensors from Japan to Alaska, verifying that the Russians had indeed tested an atomic device.

While the Soviets remained silent about RDS-1, on 23 September President Truman publicly announced the event and began an accelerated debate regarding the desirability of proceeding with development of thermonuclear weapons. Exactly two months later the US Joint Chiefs of Staff informed the Secretary of Defense that 'possession of a thermonuclear weapon by the USSR without such possession by the United States would be intolerable'.

The next Soviet atom bomb tower-test took place more than two years later. On 24 September 1951, RDS-2 was detonated close to the ground to test a tritium-boosted uranium bomb with a levitated core delivering a yield of 38KT; but the most significant was RDS-3, a 41KT bomb, dropped by air from an altitude of 10,000m (32,800ft) and detonated at a height of 400m (122ft). Of composite uranium/plutonium design, RDS-3 had a plutonium efficiency of about 35% and was announced by the US AEC three days later.

Dropped from an Ilyushin Il-28 on 23 August 1953, the 1,200kg (2,645lb) RDS-4 was detonated at 600m (1,970ft) with a yield of 28KT. The device was a boosted fission bomb using a

created to test uranium slugs and various other materials to be used in production reactors. Several design characteristics were similar to those of the Hanford 305 test reactor in that it had a diameter of 5.8m (19ft), had a loading of 25–50 tonnes, a lattice spacing of 20cm (8in) and control rods with a diameter of 3–4cm (1.2–1.6in).

As plans matured and work facilitated the first Russian atomic test, plutonium was produced at the industrial site of

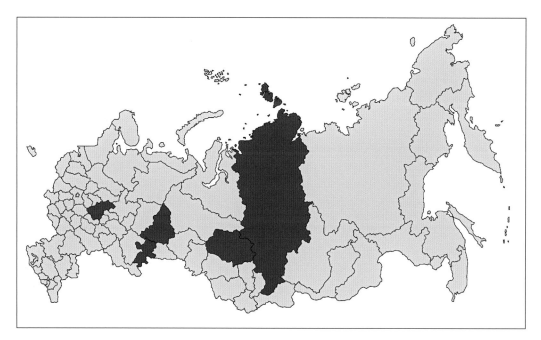

plutonium levitated core design. It was the first tactical nuclear weapon deployed by the USSR, and although it would not become operational before 1954 it would be carried by Tupolev Tu-4 and Tu-16 aircraft. This was also the first nuclear device attached to a missile, the R-5M, known in the West by its NATO code name SS-3 Shyster. The proximity of potential enemies encouraged the Russians to speed ahead with deployment of fissile devices so that they could be used to force open gaps in NATO defences or troop concentrations. But the ultimate goal was always to develop thermonuclear weapons for strategic deployments on long-range aircraft and missiles, and this had been a step in that direction.

It is now clear that the determination of the Russians to move quickly from a fissile bomb to a fusion device had a degree of seamless inevitability lacking in the decision in the United States to acquire a thermonuclear weapon. Despite this America would get there first, but a determined effort in Russia to catch up and acquire the hydrogen bomb brought an accelerated programme that would close the gap between the Soviet atom bomb and the hydrogen bomb to just six years. A genuine lack of enthusiasm to acquire the 'Super' bomb left the Americans waiting eight years.

When the Americans did test their first hydrogen bomb on 1 November 1952 it consisted of a massive test rig weighing 82 tonnes, quite incapable of being carried by

an aircraft; not before 20 May 1956 would they demonstrate an air-drop weapon, sized for operational deployment in a test known as Cherokee. By that date the Russians had demonstrated their own thermonuclear weapon – dropped by air six months before Cherokee did it for the Americans.

In the intervening period between the first American thermonuclear ground test and their first air-drop test, America's 15MT Bravo test of March 1954 had prompted the

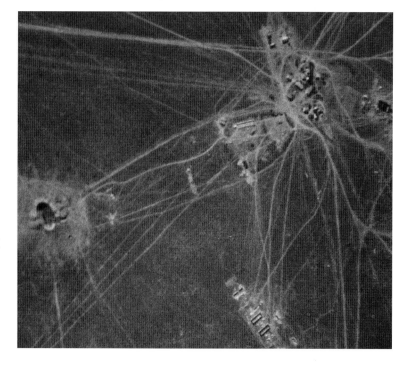

Russians to begin an intensive programme seeking a design configuration to match it, and development had not been without a lot of deliberation about the optimum route to take. Working with Andrei Sakharov, Yakov Zel'dovich proposed a two-stage device on the radiation implosion principle whereby the initial fission device would release X-radiation and trigger implosion of the fusion stage.

By the beginning of 1954, fortuitously in time for the post-Bravo acceleration of effort, Zel'dovich and Sakharov proposed their solution. Opposition came from a group that believed the spherical symmetry of the explosion in the initial stage could be focused uniformly. Sakharov led the charge to destabilise the critics and delivered a robustly sound defence of this on the grounds that the amount of energy in the initial release was so intense as to overwhelm the secondary. But in working through the problem solved in the United States by the Teller-Ulam method, Sakharov came to a completely different solution quite independent of any external help.

Under the project test designation RS-37, his innovative and unique solution was to incorporate the primary and secondary devices into the same compartment in the so-called canonical configuration that would maximise the concentration of X-rays which, because of the arrangement, would cause a fusion reaction. The detailed specification of the configuration was decided by 3 February 1955 and the RDS-37 device was delivered to Semipalatinsk. A final change occurred when it was discovered it was possible to replace the deuterium-tritium fuel with a lithium-deuterium mix, a change prompted by the published data on the Teller-Ulam tests.

In seeking an early success, Russia had already detonated a precursor device which incorporated a single-stage, fission-triggered ignition of lithium-6 deuteride in a layered design which, when detonated on 12 August 1953, delivered a yield of 400KT, which was approximately ten times the yield of bombs tested to this date. It was a conventional two-stage bomb with a deuterium-tritium secondary, and in this it was similar to the traditional line of development taken by the Americans. In fact, RDS-6 was rushed through as a fast-track compromise so as to allow the propaganda boast that Russia had tested its own hydrogen bomb. But that was not the case and the RDS-6, while powerful, could not be scaled further and was not a true hydrogen bomb at all.

The definitive Russian hydrogen bomb, and the configuration which would serve as the template for future development, was dropped by air from a Tu-95 bomber on 22 November 1955 with a yield of 1.6MT, the blast wave focused down due to an inversion layer in the atmosphere. A young girl was killed 65km (40 miles) away when a building collapsed in Kurchatov, and an intense and almost insufferable heat was felt by people 32km (20 miles) from the detonation. But the bomb had delivered – both on performance and as a political tool for the USSR, and this success was coupled to the rapid development of Russia's ICBM, the R-7. First launched on 15 May 1957 it was ahead of America's Atlas ICBM, which flew on 17 December that same year.

However, as with the Americans, it would take several years for the Russians to marry a production nuclear bomb to a mature, reliable

BELOW A reconnaissance image of Semipalatinsk-16, one of the most secret areas in the history of Cold War technology and the place where new and innovative research projects were developed, not just in the range of nuclear weapons but in a broader range of physics and applications to weaponry and defence. *(David Baker)*

and credible deterrent capability, based largely on the intercontinental ballistic missile. From the Soviet viewpoint, the total encirclement of the USSR by NATO countries significantly reduced the possibility of hitting the United States with conventional aircraft, especially because of the extensive US air defence networks. Which is why long-range and intercontinental missiles were so important to the integrated deterrent strategy of the Soviet Union.

The insatiable appetite for increasingly powerful weapons for propaganda purposes, and the pressure applied by the senior political leadership, sometimes eclipsed the operational requirements of the military. The world's biggest bomb emerged from just such a demand following two years of a test moratorium that began in 1958. With stresses added to an already sour diplomatic discourse between East and West, Soviet Premier Khrushchev responded to deepening tension by resuming nuclear testing with a new series beginning on 1 September 1961.

Khrushchev's determination to show strength and Russian technology, only a few months after the Soviet Union had placed Yuri Gagarin – the first man in space – in orbit, brought about the RDS-220 detonation, known subsequently as the 'Tsar Bomba', which occurred on 30 October 1961 with an explosive yield of more than 50MT. Khrushchev had urged his scientists to produce a bomb with a yield of 100MT and to do that the physicists constructed a three-stage configuration. But instead of a uranium fusion stage tamper they replaced the tertiary stage with one made of lead, which had the effect of halving the yield but eliminating 97% of the radiation fallout, creating 1.5MT of fission instead of 51.5MT but resulting in a yield of about 57MT.

Had the original 100MT yield been expressed it would have increased by 25% the total amount of global radioactive contamination that had been produced by all the nuclear weapons tested to that date. Nevertheless, the bomb was enormous, weighing 27 tonnes and carried by the only aircraft capable of dropping such a behemoth, a specially modified Tu-95V. Moreover, a 100MT device would probably have consumed the aircraft that dropped it. The entire exercise was a propaganda move to ratchet up the tension and show indiscriminate resolve to

face down the West. It had been preceded by a 25MT test on 23 October and the propaganda claims had already started. Russian news media announced a succession of tests and the newspapers were filled with stories of ever-bigger bombs. So it was that the world was poised for a 100MT device that never came. It was enough – even for Khrushchev.

The effect of the 'Tsar Bomba' had worked its trick. The bomb had been dropped over the Novaya Zemlya test site and it had been spectacular. Released at a height of 10,300m (33,800ft) and detonated at an altitude of 4,000m (13,000ft) 6min 8sec later, the fireball had attempted to touch the ground, but the shock wave prevented this and it swelled to a height of 8,000m (26,250ft). It was observed 1,000km (620 miles) away by the State Commission, a distance necessitated because the blast alone would cause third degree burns at a distance of 100km (62 miles).

All the buildings, including those made of brick, were totally destroyed in a village 55km (34 miles) away and a spectator wearing goggles felt the heat from the thermal energy 270km (170 miles) distant. The shock wave could be seen visibly propagating through the air at a range of 700km (430 miles) and window panes were shattered at 900km (530 miles). By

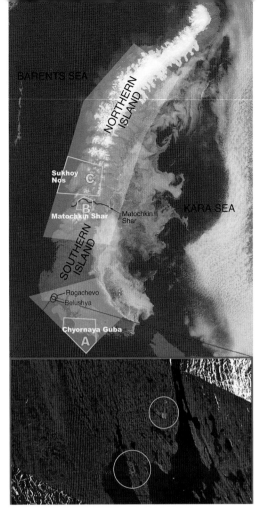

lensing through the atmosphere, windows were broken in Finland and Norway and the seismic shock wave that hit the ground went round the world three times and recorded a seismic disturbance of 5.25.

Fuel production

With a long-term investment in nuclear weapons, production of weapons-grade uranium and plutonium was a vital part of supplying the manufacturing chain all the way to type-test and operational deployment. Eventually Russia would set up four plutonium production centres: at Kyshtym, 15km (9 miles) from the city on the east side of the Urals in Chelyabinsk province; at Dodonovo, on the Yenisey River north-east of Krasnoyarsk in Siberia; a facility near Tomsk; and Beloyarskiy, near Sverdlovsk north of Kyshtym. The reactors at Kyshtym and Dodonovo were dedicated plutonium production facilities but the other two doubled for power production.

The Kyshtym facility was developed between 1945 and 1948 using 70,000 inmates from 12 labour camps and became known first as the Chelyabinsk-40 site and then Chelyabinsk-65, extending over an area of approximately 2,700km^2 (1,042 miles^2, or 270,000ha). In fact, the facility was in the city of Ozyorsk, a closed place under the Communist regime. A significant accident occurred on 29 September 1957 when 80 tonnes of fission products were released through an explosion with a force of 100 tonnes of TNT, directly contaminating an area of up to 1,000km^2 (386 miles^2, or 100,000ha).

Up to 10,000 people from 22 villages were evacuated and resettled, some only after two years, with redirection of the water systems to isolate the area from Lake Kyzyltash. The radioactive cloud of caesium-137 and strontium-90 brought about the deaths of between 6,000 and 8,000 people as a result of the contamination, figures derived from norms of mortality from cancer before the accident compared to those in the following decades. The Soviet authorities disguised the contaminated area by designating it as a special nature reserve where people were prohibited for the preservation of a 'natural ecology'.

Large quantities of plutonium were produced in all these reactors, supplying the rapidly expanding inventory of Soviet nuclear weapons, including bombs, warheads, mines and shells. Numerous commercial power reactors too were eventually tuned to the production of plutonium

for weapons. The estimated production quantities are varied simply because so many of the records have either been destroyed or are still a secret known only to the highest authorities. But a general idea can be gained from a conservative calculation based on known facilities and verified output.

Assuming 0.86g/Mwd (grams per megawatt-day-thermal) for 6% of Pu-240, in 1958 some 7kg (15.4lb) was produced, rising to 150kg (330.7lb) by 1961 and 1,800kg (3,970lb) by 1965. Less than a decade later annual production topped 1,300kg (2,866lb), reaching 3,000kg (6,615lb) in 1982, at which point it stabilised at that annual rate. By the end of the Cold War in 1991 the total cumulative mass of Pu-240 reached 51,000 tonnes. The end use was to provide the USSR with a gradually

expanding stockpile to catch up and eventually exceed that of the United States, exceeding 30,000 nuclear weapons by the late 1970s to the 28,000 of the United States.

Tritium production for thermonuclear weapons really began after news of the American hydrogen bomb but Kurchatov had already established a working group, including Sakharov and Igor Tamm, to have the first isotope reactor working by 1952. With the availability of heavy-water reactors by 1951, the Russians had an efficient method of producing tritium. Most of the subsequent designs used graphite-moderated types.

Uranium production exploited one of the richest continental depositories of this material anywhere in the world and Russian geological literature abounds with exotic descriptions of

ABOVE Extraordinarily efficient and with extensive countermeasures, a turboprop-powered Tu-95 Bear H is escorted during a fly-by off the coast of Scotland, a political tool every bit as viable today as it was 50 years ago. *(RAF)*

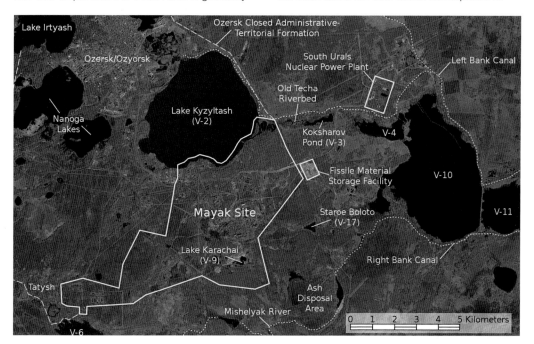

LEFT This satellite image of Chelyabinsk shows the Mayak processing facilities and the location of fissile storage and ash disposal facilities. *(Via David Baker)*

ABOVE Nicknamed 'Tsar Bomba', the RDS-220 thermonuclear device was detonated on 30 October 1961 and delivered a yield of 57MT, the most powerful bomb tested. It was dropped by a Tu-95 on the tip of Severny Island, part of the Novaya Zemlya archipelago. *(Via David Baker)*

(UF$_4$) through reaction with fluoride and reduction to a metal for weapons, reactor fuel or for conversion to gaseous hexafluoride (UF$_6$) for enrichment to the U-235 isotope. Like the Manhattan scientists, Russian physicists evaluated the three methods of enrichment: gaseous diffusion, electromagnetic separation and gas centrifuge.

While Britain and America received the assistance of several noted German and Eastern European physicists fleeing Nazi oppression before the war, Soviet research gained the benefit to a lesser degree from German scientists at the end of the conflict, among them Professor Adolf Thiessen. In 1948 Thiessen developed a laboratory-model gaseous diffusion barrier and this was chosen as the template for a mass-production plant at Elektroskol, south of Noginsk near Moscow.

Procurement of material was under way by 1949, but this was just one of several methods for isotope separation and by 1950 it had been decided to put all resources into the gaseous diffusion method. The first cascade was built at Kefirstadt, later named Verkhniy-Neyvinskiy, but corrosion within the barrier prevented the production of weapons-grade uranium of a purity greater than 90%.

In a parallel exercise, Dr Max Steenbeck,

every conceivable type of uranium possible. They include deposits associated with ores and albitites in Precambrian metamorphic rocks and phosphates in clays. The methods of exploration and mining involved the typical approaches used elsewhere in the world, including in-situ leaching at the surface, open-pit extraction of low-grade ores and underground mining to obtain the high-grade deposits in veins 200m (656ft) or more below the surface.

The stages in obtaining enriched uranium followed established procedures, with extraction of uranium oxide (U$_3$O$_8$) at the mining site, followed by conversion to uranium tetrafluoride

RIGHT On 29 September 1957 a major accident occurred at the Kyshtym plutonium production facility, leaking caesium-137 and strontium-90 over a very wide area, as traced on this map of the region. *(David Baker)*

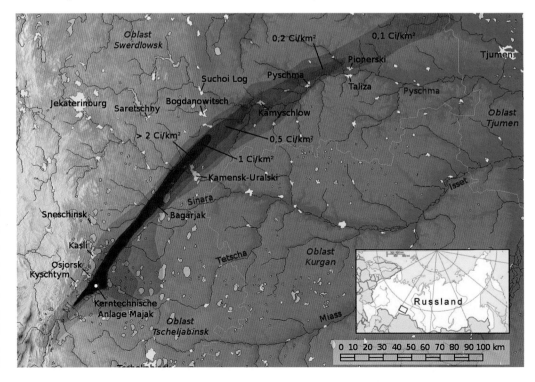

formerly of the Siemens Company in Germany, helped the Russians to develop a gas centrifuge research facility at Sinop on the Black Sea coast. While Steenbeck was building a centrifuge topping plant to enrich the output of the diffusion cascade from 50% to 90%, Thiessen was working to correct the corrosion problem with his diffusion barrier, which was solved by 1951. It was from this development that the first nuclear device was tested using highly enriched uranium.

Nevertheless, despite success with gaseous diffusion methods they persisted with research on the gas centrifuge concept and the topping plant was redirected to develop a centrifuge plant capable of taking natural uranium to 96% U-235 at the rate of 1kg (2.2lb) a day. This work was transferred to the Kirov plant in Leningrad on 15 September 1951 and over the next two years the Germans were gradually eased out of the project so that by September 1953 the Russians had gained total control. From that date virtually all weapons-grade production was through the gaseous diffusion method, although several other methods were researched including photochemical technology using lasers.

Only slowly did the West become aware of the scale and the magnitude of Soviet uranium production and accurate figures are difficult to come by, even today. At the end of the 1970s it was believed that a single gaseous diffusion plant, perhaps the one at Krasnoyarsk, was producing 7–10 million kg separative work units (SWUs) per year, equal to about half the total US peak output of 16.5 million kg SWUs. It was believed the Soviets had three plants of the Krasnoyarsk capacity.

By the early 1980s the USSR had approximately three million SWUs/yr which were available for export through its agency Techsnabexport and that it had contracts for exporting 0.6–3 million SWUs/yr. Intelligence analysis indicates that in the 30 years beginning in 1946 the Soviets built a 200,000-tonne stockpile of uranium which would have been sufficient to produce a total 600–700MT of weapons-grade uranium using gaseous diffusion. Further analysis and calculation of power plant electrical output indicated a civilian uptake of only 10%, the rest going to weapons production.

One of the major weapons-grade plutonium reactor complexes was the Zheleznogorsk plant, which operated from the late 1950s and continued to produce plutonium after 1993. Under the terms of the 1997 Plutonium Reactor Agreement between America and Russia the plutonium was not used to produce weapons but continued to add to the Russian stockpile. In April 2010 the plant was in the closing stages of final shutdown after 52 years of production. The only two other remaining plutonium reactors, at Seversk, had already closed.

Russian tests

Between 29 August 1949 and 1 May 1991 the Soviet Union conducted 727 nuclear tests including peaceful nuclear explosions (PNEs), but like the US, Russian scientists frequently conducted more than one detonation at a time. Including all detonations, there were 981 shots of which 156 were PNEs. Some 114 were safety tests, weapon-effects tests, fundamental physics tests and the notorious Totsk test of 14 September 1954 in which

ABOVE The R-2 (NATO name SS-2 Sibling) was a developed version of the R-1, Russia's copy of the German V-2 and a source of test and experimentation immediately after the Second World War. This missile adorns the city of Korolev in the Moscow oblast. *(David Baker)*

ABOVE Designed and developed as an improvement on the German V-2, the SS-1b Scud is arguably the most ubiquitous missile developed by the former Soviet Union. Capable of throwing an 80KT warhead a distance of 300km (180 miles), it entered service in 1957 and has been exported to many countries around the world for launching conventional warheads. *(David Baker)*

ABOVE RIGHT Introduced in 1956 and designated in the West as SS-3, the R-5 Pobeda could carry a variety of different nuclear warheads and had a range of 1,200km (750 miles), insufficient to reach the United States. It nevertheless remained in service until 1967. *(David Baker)*

nuclear detonations were conducted during a huge military exercise without any warning being given to the civilian population.

Russian test sites were located primarily at Semipalatinsk and Novaya Zemlya, with other sites in the Ukraine, Uzbekistan and Turkmenistan. The total collective yield of all Soviet/Russian nuclear tests at these various locations amounted to 296,837MT, representing 54.9% of all nuclear testing by all countries around the world.

Soviet nuclear tests had been conducted on a relatively spasmodic basis between 1949 and 1957 but on 1 September 1961 an intensive series of testing resumed, largely in response to a worsening crisis in international relations, events associated with this period being the construction of the Berlin Wall through to the Cuban Missile Crisis of October 1962 – two years in which 78 tests took place.

It was during the 1961 sequence that the 57MT 'Tsar Bomba' was detonated, which brought international reactions that began to turn against the USSR. The bombast from Khrushchev only served to alienate potentially sympathetic countries and the mood brought about a series of agreements that changed the way the superpowers approached the almost indiscriminate proliferation of nuclear testing. But attention to these disturbing consequences had already set actions in motion that would begin a transformation in international recognition of a problem that was rapidly escalating toward a global catastrophe, driven by radioactive contamination of the environment.

The general suspicion among the populations of the non-communist world began to grow, a sense that science had really not produced solutions to the indiscriminate use of nuclear weapons for tests in the atmosphere.

The US nuclear test in March 1954 had been twice the yield expected and caused fallout far beyond the predicted zones. When a Japanese fishing vessel was contaminated and its crew received radiation sickness, as well as occupants of a nearby atoll, global outrage reached a peak. In a similar incident, radioactive rain from a Soviet test fell on Japan.

Knowledge of the effects of radiation grew quickly, and the alarming reality emerged that there had been far too little attention given to the consequences of the atmospheric tests. Global surveys soon showed that no place on Earth was immune from the effects and the consequences for genetic damage began to be defined. Pressure groups began to mobilise and protests urged action to stop the tests, with appeals from the West to 'closed' societies in communist countries to support a ban on nuclear weapons tests.

Paradoxically, at least for the general assumption regarding the approach taken by the USSR, it was from Russia that some of the most focused appeals for a moratorium originated. The public got involved when, on 10 May 1955, the Soviets included discontinuance of general weapons trials in its proposals for ending nuclear tests. Later that year in the UN General Assembly when the Soviets advocated a separate test ban, the US, Britain and France spent the next three years asserting that agreement would be contingent on progress in other matters of arms control. With the exception of France, they dropped the linkage in January 1959, by which time the Soviets had decided to halt nuclear testing, effective from 31 March 1958.

Several times the Russians had offered a complete ban under the direction of an international authority. On 14 June 1957 they offered test ban proposals that included international control, with very general proposals for supervisory groups established on a basis of reciprocity on the territories of the then three nuclear powers (US, USSR and the UK). A complete moratorium on testing was proposed by Moscow. But the proposal of March 1958 was new and far reaching. The Americans dismissed this Russian appeal for all nuclear powers to desist and the US began its Operation Hardtack I series of nuclear testing on 28 April that year.

Ramifications of the appeal from Russia for

ABOVE Infamous as one of the missiles deployed to Cuba in 1962, the R-12 (SS-4 Sandal) had a range of 2,080km (1,290 miles) and could carry a 2–3MT warhead. In service from March 1959, it was withdrawn from use in 1993 after more than 2,335 had been produced. Variants were exploited as satellite launchers and it was the first rocket to send a payload into space following the R-7, its first launch in that role taking place on 16 March 1962.
(David Baker)

an immediate suspension of tests reverberated when it challenged the other two nuclear powers to go 'to their parliaments' and seek approval for a similar restraint. But Khrushchev reserved the right to resume testing if the other two powers failed to comply. At a conference in Geneva during July and August, Russia, the US, the UK, France, Canada, Poland, Czechoslovakia and Romania discussed and agreed a proposed network of 170–180 land-based control posts and ten shipborne stations as well as special aircraft flights. But the US and the UK had continued their tests and the Russians resumed on 20 September but halted again on 3 November.

The concern that the Russian unilateral moratorium would result in a universal ban hastened the British nuclear programme for fear that tests would be outlawed before it had its own bomb perfected. But throughout 1958 the Eisenhower administration gradually changed its mood and agreed to open discussions about the technicalities involved in a test ban and associated verification agreement, meetings for which dragged on through 1959 and 1960.

At the Vienna summit of March 1961, President Kennedy and Premier Khrushchev met face to face for detailed discussions. Khrushchev had been deeply concerned about

the threat of nuclear war. While determined to match military and political opposition from countries he saw as enemies, he believed that nuclear weapons could never be used in war, that they constituted a real risk of self-destruction and that the defence of the USSR was better off without them. In this he faced vehement opposition from the military and from several of his political advisers, but in that confrontation he was not alone.

President Kennedy was equally opposed to unconstrained expansion of nuclear warfighting capabilities and made a test ban treaty an early objective of his administration. But the Vienna meeting presented little opportunity for a way out and the two leaders withdrew without even a compromise. In the US, the military and some political advisers to the administration warned the President not to go so far as to disarm America. Elements in the Department of Defense had long cautioned against trusting the Soviets and began to sow rumours of Russian cheating by conducting clandestine tests. In August 1961 the Soviets resumed testing and a difficult year got worse when the Soviets threw a wall across the former German capital, isolating the East Berliners from West Berlin.

The desire for a solution to nuclear testing in the atmosphere occupied the United Nations

BELOW Serving with the strategic rocket forces between 1962 and the early 1970s, the R-14 (SS-5 Skean) IRBM carried a 1–2MT warhead a distance of 3,700km (2,300 miles) and had an extended life as a satellite launcher. *(David Baker)*

RIGHT Josef Stalin made nuclear weapons and their delivery by rocket a cornerstone of his deterrent strategy, but not until the regime of Premier Khrushchev did the ICBM become a reality. In 1957, under Sergei Korolev, the R-7 made its first flight on 15 May 1957, and was used to place the world's first satellite in orbit on 4 October that year. *(Herberto Arribas Abato)*

FAR RIGHT The SS-7 was Russia's first successful ICBM, delivering a warhead with a yield of 5–6MT across a maximum range of 11,000km (6,800 miles). It used storable propellants, entered service in 1961 and was retired in 1976. *(David Baker)*

at the Conference on the Discontinuance of Nuclear Weapon Tests in Geneva, almost continuously between 31 October 1958 and 29 January 1962. Following an adjournment of these discussions, the newly formed Eighteen-Nation Disarmament Conference (ENDC) began meeting in March, involving the newly created Arms Control & Disarmament Agency (ACDA). Certain discrepancies were tidied up and the US, the UK and the USSR began three-power talks that culminated in an agreement on 10 June 1963 to hold talks in Moscow.

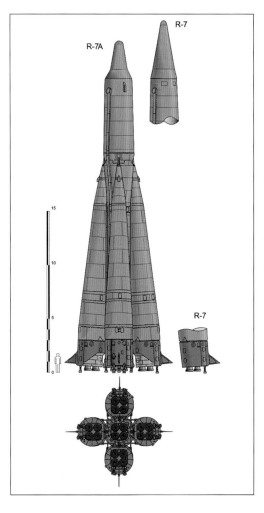

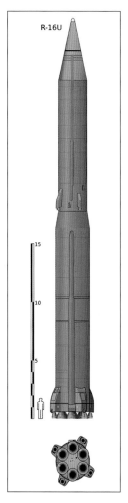

LEFT When it appeared in 1974, the R-26 (NATO name SS-9 Scarp) was the first Soviet ICBM known to be capable of carrying three separate warheads on a MIRV (Multiple Independently Targetable Re-entry Vehicle) platform. It had a range of up to 16,000km (10,000 miles), depending on the weapon load, and was a highly survivable system due to the hardened silos from which it was launched. *(Novosti)*

The three-power meetings began on 15 July and the Partial Test Ban Treaty (PTBT) banning all atmospheric testing was negotiated over the next ten days, formally signed in Moscow on 5 August 1963. On 24 September the US Senate ratified the treaty and it was signed by President Kennedy on 7 October 1963. It entered into force on 10 October. The last atmospheric test by the United States had been conducted on 9 June and by the Russians on 25 December 1962. The agreement may have been influenced by the dramatic superpower confrontation over Cuba in October 1962, but whatever the reason a major step had been taken to improve relations between the superpowers and to stop adding radioactive materials to the atmosphere.

Five years later, in 1968, the Nuclear Non-Proliferation Treaty (NPT) was signed to halt the spread of nuclear weapons. It entered into force in 1970, with more countries signing up than to any other arms limitation or controls agreement on record. As of the beginning of 2017, 191 states had become signatories to an agreement that pledges to promote the peaceful use of nuclear energy and for the powers with nuclear weapons to work toward their elimination. Nuclear weapon states are defined as those who had tested a nuclear device before 1 January 1967, which included the US, the USSR, the UK, France and China. India, Pakistan and North Korea are known to have carried out tests or to have deployed nuclear weapons operationally, with Israel refusing to confirm or deny that is has them.

In 1974 the Threshold Test Ban Treaty came into force, banning underground testing of weapons with a yield exceeding 150KT, and under the 1976 Peaceful Nuclear Explosions Treaty the US and Russia agreed to halt testing of devices associated with peaceful nuclear

LEFT Seen here on its 9P120 launcher, the 9M76 theatre ballistic missile was designated SS-12 Scaleboard when it was first observed, the missile having a range of 900km (500 miles) with a 500KT warhead. *(David Baker)*

explosions above 150KT. Negotiations between the US, the UK and the Soviet Union resumed in October 1977 but little progress was made for the next decade or so. When Ronald Reagan became President in 1981 he sought both expansive discussions at lower levels to sound out the possibility of doing a deal on stopping all tests while stiffening US military potential and inaugurating a wide range of new weapon systems.

In 1985 Soviet Premier Mikhail Gorbachev announced a unilateral test moratorium, and that effectively brought an end to Russian nuclear weapons tests of any kind. Reagan consolidated his intention to secure a Comprehensive Test Ban Treaty (CTBT) and talks resumed in November 1987, with an agreement the following month to conduct investigations into how verification of compliance could be secured for underground testing. Negotiations continued through to the end of the Cold War in 1991 and in 1993 negotiations for a CTBT began, supported by the United Nations.

It was finally opened for signature at the UN on 24 September 1996 at which point it secured 71 states, five of which were nuclear powers. By the end of 2016 a total of 166 states had signed in, with a further 17 signed but not ratified. Unfortunately China, Egypt, Iran, Israel and the US have not yet ratified the treaty, and India, North Korea and Pakistan have not signed it. The last three countries have carried out nuclear tests since the treaty opened for signature.

The Russian stockpile

From very early in the nuclear age Russia sought to acquire a deterrent primarily based on land-sited missiles supplemented to a lesser degree by succeeding generations of submarine-launched ballistic missiles (SLBMs) and by manned long-range bombers. Because Russia is a vast land mass it can conceal its ballistic missile sites deep in uninhabited parts of the country, while the United States, as a major maritime power, quickly sought to move its primary deterrent out to sea on SLBMs.

In terms of weapons and delivery systems, the catalogue of expansive Russian nuclear hardware has been referenced widely in

LEFT Faced with a coalition of NATO forces, backed up by the United States, the Russians made early strides in the development of theatre and battlefield nuclear delivery systems, including this RT-15 (SS-14) operational from the mid-1960s and capable of throwing a 1MT warhead across a range of up to 2,500km (1,600 miles). Deployed on mobile units, from Eastern Europe it could have struck targets as far west as the Atlantic coast of Ireland. *(Novosti)*

LEFT One of the more potent threats to a stable deterrent policy was the RSD-10 (SS-20 Saber), deployed in Eastern Europe from 1976. Road-mobile, this IRBM had a range of 5,500km (3,400 miles) but did not come into the strategic arms talks between the US and the USSR. Equipped initially with a single 1MT warhead, later versions carried three warheads of 150KT. It alone was responsible for the NATO deployment of Gryphon cruise missiles and the Pershing 2 as a counter during the 1980s. *(Via David Baker)*

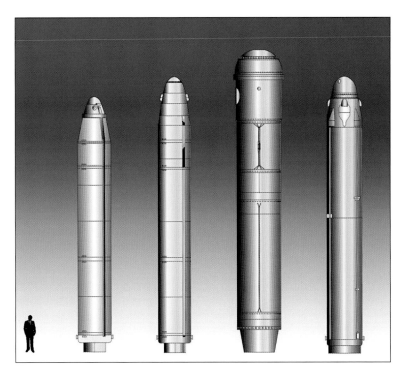

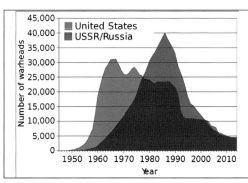

ABOVE The massive development and production effort launched by the United States immediately after the Second World War saw an overwhelming advantage in numerical superiority for US nuclear stockpiles until the mid-1970s, when Russia overtook the West. Arms agreements and expediency over budgets and *realpolitik* have since fuelled a significant decrease in total numbers deployed by the major powers. *(David Baker)*

ABOVE Russia's nuclear weapons were taken to sea on ballistic missiles in a submarine-based programme that began in 1955 at the northern city of Severomorsk in the Murmansk Oblast. Here to scale are (left to right) Russia's SS-N-8 with a range of 7,700km (4,785 miles), SS-N-18 with a range of 6,500km (4,040 miles) and the SS-N-20 and SS-N-23, each with a range of 8,300km (5,150 miles). *(David Baker)*

publications and on the Internet but the general trend for Russia has been the development and survival of ever more capable ICBMs. Deployed on exposed launch pads in the early days of ballistic missiles, ever more secure means of containing them in silos protected from all but a direct hit by a nuclear weapon has been acquired in the decades since.

Moreover, with experience from the Second World War the Soviet Union maintained a very large standing army throughout the period of the Cold War and consolidated those forces with a wide range of tactical, battlefield and regional nuclear weapons and their delivery systems. These became a focus of attention for NATO and its European members, mindful that any war between them and the Warsaw Pact forces would very quickly turn Europe into a seething furnace of atomic war and irradiation.

During the first 30 years after the end of the Second World War the US maintained a significant lead in the number of nuclear weapons in the stockpile. In the mid-1970s that situation was reversed and while the US began a slow decline in the numbers of its warheads, and reduced the yield of most weapons to a level where they could be more effectively managed in the prosecution of an actual conflict, Russia increased its stockpile, reaching a peak of about 40,000 warheads

by the mid-1980s, at which time the US had around 22,000. It was this imbalance that prompted President Reagan to engage in a major reconsolidation of conventional forces.

During the last few years of the USSR, at the end of the 1980s, the number of Soviet warheads began to decline until today there is a greater level of parity than there has been at any time since the Second World War. But the distribution is very different: the US still favours SLBMs over ICBMs, quite the reverse of the policy of the Kremlin, which favours land-based delivery platforms.

Today the balance between nuclear superpowers is much more evenly set, and less reliant on the use of such weapons to achieve political or military gain. Like America, Russia's strategic nuclear forces are primarily based on missiles in fixed silos or on road-mobile launchers and there is considerable development under way in each category. Russia still retains a viable SLBM fleet and the manned bomber is consolidated with upgrades to existing Cold War types. But the threat of placing nuclear missiles on client states at the borders of NATO still resonates strongly with leaders of minor countries adjacent to the Russian border.

In 2017 (at the time of writing) Russia has approximately 7,000 nuclear weapons. Of this total 4,300 are operationally assigned to strategic delivery systems, of which 1,950 are deployed on ballistic missiles and at heavy bomber bases and 500 are in storage together with 1,850 non-strategic warheads. About 2,700 are retired but intact and awaiting dismantlement. Russia is fully compliant with international agreements on launcher numbers and warhead totals but is engaged, like the US, in a force modernisation structure that will considerably enhance operational capabilities.

To conform to the New Strategic Arms Reduction Treaty (START) limits, modernisation plans should stabilise the Russian force at 500 strategic launchers with a total of 2,400 assigned warheads. Currently Russia has 316 ICBMs carrying a total of 1,076 warheads out of a declared total of 847 deployed and non-deployed launchers, which includes ICBMs, SLBMs and bombers. The precise balance between deployed and non-deployed is uncertain as that information is classified by agreement within the Russian and US defence authorities.

There are nearly 800 warheads carried on 176 SLBMs deployed with 11 submarines and about 50 deployed strategic bombers with, theoretically, a total of more than 600 nuclear-tipped cruise missiles. Both Tu-95MS and Tu-160 strategic bombers carry subsonic KH-55SM cruise missiles exclusively, which has a range of up to 3,000km (1,860 miles), and no longer carry gravity bombs. A successor, the very stealthy Kh-102, has a range of up to 5,500km (3,417 miles) and is predicted to be in service by 2023.

Like the United States, Russia is silent on the number or types of non-strategic theatre and tactical nuclear weapons in its arsenal but it is believed to number almost 2,000 and is slowly declining. The Kremlin is committed to maintain a balance equivalent in this area to the nuclear forces of the US, the UK and France combined – the nuclear powers of NATO. But there are signs that Russia is deploying the SS-26, a pseudo intermediate-range ground-launched cruise missile in violation of the Intermediate-Range Nuclear Forces (INF) treaty, an agreement on shorter range missiles. It carries two missiles with a range of 300km (186 miles). Throughout its deployments it seems clear that Russia relies on nuclear weapons to defend against overwhelming conventional forces, a position not far removed from its stance during the Cold War.

ABOVE Escorted by a Sukhoi-Su30, a supersonic Tu-160 Blackjack long-range bomber drops a KH-55SM nuclear-tipped cruise missile, part of Russia's modernisation programme for delivery platforms, some of which are now capable of targeting any location on Earth from its own territory. *(Russian Air Force)*

BELOW With an operational range of 11,000km (6,800 miles), the RT-2PM2 Topol-M strategic ballistic missile is a dramatic demonstration of Russia's intention to maintain the highest level of nuclear delivery systems necessary for an effective deterrent. It is being replaced and supplemented by the road-mobile RS-24 Yars. *(David Baker)*

Chapter Four

The British bomb

Determined to remain at the "top table" and to deter conflict by threatening nuclear annihilation on aggressive powers, the UK began development of an independent nuclear weapon and to build forces capable of immediate and overwhelming response to armed threats.

OPPOSITE Custodian to 16 Trident submarine-launched ballistic missiles, HMS *Vengeance* returns to its naval base at Faslane on the Clyde following an at-sea training exercise. *(Royal Navy)*

ABOVE The agreement to share development of the American atom bomb had been drawn up by US President Franklin D. Roosevelt and British Prime Minister Winston S. Churchill, which had no binding commitment in law and was retired with Roosevelt's death on 21 April 1945.
(Via David Baker)

RIGHT William Penney (1909–91) was a vital part of the British contingent that went to Los Alamos to work on the Manhattan Project, but after the war he led the effort to build a British bomb and was made Rector of Imperial College, London, which built and named after him the William Penney Laboratory.
(David Baker)

Until President Roosevelt decided in 1942 to start the development of an atomic bomb in the United States, the UK was itself within reach of making a fission device but lacked the material and financial resources to make it happen. When the war ended in September 1945 the UK understood that it had full access to US developments with nuclear bombs, emphatically endorsed by the Hyde Park Aide-Mémoire of 19 September 1944. Only a very small circle around Churchill, probably no more than seven ministers, knew of the existence of

the bomb programme until Hiroshima. After the completion of work at Los Alamos the British scientists went home. But, in the words of Sir John Cockcroft, the British had 'an almost complete knowledge of technology' associated with the bomb.

To the surprise of the UK, and its Prime Minister Clement Attlee, who had not been privy to knowledge about the Manhattan Project and Britain's involvement in it when he was deputy to Churchill, the US Atomic Energy Act of 1 August 1946 forbade any transfer of technological knowledge or scientific information related to atomic weapons. Britain had been unable to develop a bomb of its own during the war but now, denied any cooperation with the United States, it was felt imperative for it to resume that development and to acquire an independent atomic bomb designed and built in the UK.

The view had been prevalent toward the end of the war when Churchill sent a telegram to President Truman on 12 May 1945, four days after the German surrender, asserting that he was 'profoundly concerned about the European situation'. Moreover, he said, with France weakened by war and the Americans about to leave the Continent for Japan, the unknown situation regarding Russia was made worse because 'An iron curtain is drawn upon their front. We do not know what is going on behind…'

Fears of Soviet aggression mounted and on 1 January 1946 the Chiefs of Staff issued a report to Prime Minister Attlee stating that 'we require as quickly as possible the greatest capacity to make atomic bombs that economic factors and the supply of raw materials will allow'. They went on to assert that 'the best method of defence against the atomic bomb is likely to be the deterrent effect that the possession of the means of retaliation would have on a potential aggressor'.

An Air Staff Operational Requirement (OR) 1001 was issued on 9 August 1946 calling for rapid development of atomic weapons, and on 8 January 1947 a small committee of government members decided that Britain should have its own bomb. The committee, known as Gen 163, had taken its stance from the top-secret Gen 75 group that first sat in August 1945 to determine UK policy on atomic matters, including energy. The UK felt that to possess such a weapon alongside the United States would give it greater status at the United Nations, where it was one of the five permanent members, and that it would help maintain its standing within the international arena of diplomatic affairs.

As a side-note, it is interesting to observe that these five members (US, UK, Russia, France and China) were the only countries to openly acquire nuclear weapons until their acquisition by India in 1974 and Pakistan in 1998. The 'Top Five' dominated Cold War political and diplomatic control of global matters, and the possession of nuclear weapons was certainly an important part of maintaining an influence in international affairs disproportionate to their financial, manufacturing or trade status – criteria that had held sway in such rankings since the time of the industrial age in the 18th century.

This was a time when former colonies of Britain, France, Belgium and the Netherlands sought independence. Reluctance to afford that liberation was met with violence and insurrection in South-East Asia and the Far East. The Attlee government was well aware that the British Empire was at an end and that its parlous financial situation would not support military operations as they had between the world wars, when Britain supported lands and territories around the globe.

LEFT Air Chief Marshal Sir Charles Portal relinquished his role as former chief of the Air Staff to manage a group within the Ministry of Supply that put together the structures necessary for providing the British government with an independent nuclear weapon that could be delivered by the Royal Air Force. *(RAF)*

The concept of the atomic bomb being a deterrent grew rapidly, along with the realisation that it was more cost effective to develop and deploy nuclear weapons than fund large ground, sea and air forces. The nuclear bomb was considered a cheap option compared to the rest. Moreover, it would underpin a technology capable of supporting a nuclear power industry that Britain had been intent upon acquiring since before the war, when it had opted to put that on hold while its scientists went to the US to work on the American bomb.

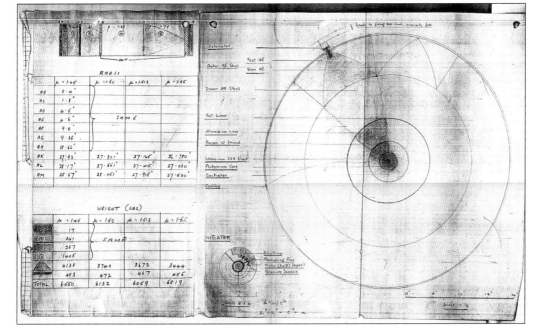

LEFT The first drawings of the British atomic bomb, December 1947, supervised by William Penney. *(Via David Baker)*

In February 1947 the UK Atomic Energy Commission was established to replace Gen 75. It would meet generally two or three times a year and this resulted in a determined effort to develop nuclear capability, which at the time was generally considered to be at least five years away from producing a working bomb. Three men were selected in early 1947 to run the British programme: John Cockcroft was in charge of a research establishment to be built at Harwell; Christopher Hinton would manage the design, construction and operation of plants to produce the fissile material; and William Penney would be in charge of designing, making and testing the bomb.

Born in 1909, William Penney was arguably one of the most important people involved in the development of a British atomic bomb, due largely to his role as head of the British scientific contingent that went to the United States to contribute to the Manhattan Project. Penney tasked all his British colleagues to keep notes of their activities and to observe all activity around them and maintain diaries recording stages in the development of the American bomb. When he returned to the UK at the end of the war one of Penney's most valued informants was Klaus Fuchs, who was to be uncovered later as a prolific purveyor of information to the Soviet Union as well!

At the time of the decision to build a bomb, Penney was Chief Superintendent of Armaments Research based at Fort Halstead, north of Sevenoaks in Kent. It was there that the initial work on what was euphemistically known as High Explosives Research was conducted, Penney's small team operating in a specially fenced-off area within which mechanical, electrical and theoretical physics departments were established. Penney's principal task was to work on the spherical ball containing high explosives that would produce the implosion wave acting on the plutonium core which would cause a nuclear explosion.

Development of the Mk 1 was carried out in close cooperation with the RAF due to its imminent deployment on completion and Squadron Leader John Rowlands was responsible for the service end of the operation. His job was to ensure it conformed to OR 1001 and to prepare all the necessary procedures using a team of nine serving RAF officers all highly qualified in mathematics, science and engineering, all of them co-located with Penney's incarcerated team.

But public acknowledgement that this was under way came in a circuitous manner, redolent of the extreme secrecy which pervaded the entire nuclear weapons programme – and still does. In answering a question in the House of Commons on 12 May 1948 about whether he could provide any information on the development of atomic weapons, the Minister of Defence (A.V. Alexander) said that he did not think it would 'be in the public interest to do that'. In other words, 'We are but I can't tell you about it.'

There is an interesting adjunct to the story of the British bomb. In the spring of 1948, at a time when the UK was excluded from access to American bomb science and engineering technology, the military on both sides of the Atlantic recognised that the deterioration in world affairs meant that the reconstruction of the old wartime alliance was a vital imperative for combating communist aggression. The dramatic communist coup in Czechoslovakia on 25 February 1948 appeared to endorse that gloomy view of the future. The preceding winter months had seen the only sensible assessment of what it would take to deter the Soviet Union from aggressive military action against the West, carried

BELOW The Atomic Weapons Research Establishment, Aldermaston, where Britain's nuclear warheads were assembled, was established on 1 April 1950 with Sir William Penney as its first director. It is now known as the Atomic Weapons Establishment, operated as a public/private partnership. *(AWE)*

out by the US Air Force Directorate of Intelligence.

Coming just a few months after the Air Force achieved independence from the Army as part of the Defense Authorisation Act, it was the first determination of a militarily sound force assessment and concluded that 70 atomic bombs would be sufficient to devastate the USSR. But this was not the only rational assessment and the British decided that a total inventory of 600 bombs, of which 200 would be provided by the UK, would be necessary by 1957 to deter the Soviets.

What was to unfold in the UK over the following few years startled and alarmed the Americans, when the US was informed in March 1948 that work had been under way for 15 months on a special section on the Ministry of Supply led by Lord Charles Portal, former chief of the air staff during 1940–45. Imagining they had cut off the possibility of a British bomb with the McMahon Act, the US Atomic Energy Commission moved to tighten the political constraints while the US military worked in whatever way they could to untie the knot.

The size of the British effort was determined to some degree by the scale of the requirement and in 1948 the government figure of 200 bombs sounded about right. This was considered the proportionate balance between a runaway escalation in stockpiling and the not insignificant cost of developing and maintaining a deterrent force. In this the UK was the first nuclear power in the world to orchestrate the development of materials production, bomb design and manufacture, stockpile production and, essential for delivery, a means of carrying the bomb to the enemy. That last task was given to RAF Bomber Command and its V-bomber force.

Also known, somewhat euphemistically, as the Atomic Energy Research Establishment, Harwell was to be built adjacent to an existing RAF airfield, 19km (12 miles) south of Oxford. During the war years it had been envisaged as the home of Britain's post-war nuclear power industry. Few people worked at Harwell until 1950, among them Klaus Fuchs.

Fissile material would be produced at Risley, south-west of Manchester, and it became a landmark of exemplary practice in diversity, complexity and pioneering excellence matched

to a high level of efficiency. It was to produce both uranium and plutonium materials, which called for a large experimental reactor (Bepo – for British Experimental Pile-0) at Harwell, two air-cooled reactors at Sellafield in Cumbria, a chemical separation plant at Sellafield where plutonium would be recovered from irradiated fuel rods, and a plant for producing uranium fuel rods at Springfields near Preston.

Similar in design to the Americans' X-10 at Oak Ridge, Bepo achieved its required criticality on 3 July 1948, followed by the two air-cooled reactors at Sellafield – renamed Windscale – in October 1950 and June 1951. They would operate for seven years before being shut down in October 1957 after a serious fire, replaced by four reactors built at Calder Hall on the opposite side of the River Calder. A chemical separation plant was also built, but instead of the precipitation process used by the Americans this would employ the Butex process of solvent extraction, and the first active material was fed in on 25 February 1952. Additional plants, for plutonium purification and finishing the plutonium into metal, were also built and the first plutonium billet was made on 31 March 1952.

Tasked with designing and building the bomb, William Penney gathered a group of 34 selected scientists at Woolwich where he outlined the requirement and explained his plan. Initial work took place at the Armament Research Establishment at Fort Halsted in Kent,

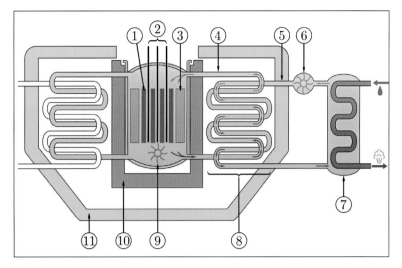

ABOVE A schematic of the Dounreay Fast Reactor (DFR), a derivative of a fast breeder reactor.

(Emoscopes)

1 Fissile Pu-239 core
2 Controls rods
3 U-238 breeder blanket
4 Primary sodium-potassium (NaK) cooling loop
5 Secondary cooling loop
6 Secondary NaK circulator
7 Secondary heat exchanger
8 Primary heat exchanger
9 Primary NaK circulator
10 Boosted graphite neutron shield
11 Radiation shield.

under the understated title of High Explosives Research Unit (HER), but it used facilities at Woolwich and at a firing range at Shoeburyness. The test range at Orfordness, Suffolk, was chosen as the location for environmental testing of warhead explosive assemblies.

Security concerns over the dispersed locations for various elements of the bomb had compromised the integration between Harwell, Risley and HER and this prompted a decision to focus weapons research at Aldermaston, Berkshire. In addition, it was decided to establish an open range at Foulness Island in the Thames estuary as an outstation for experimental work with high-explosive assemblies. Aldermaston was officially established on 1 April 1950 and the facility began processing plutonium from May 1952.

Development of the British A-bomb proceeded on a path linked to quantity production and operational deployment of the Mk 1 with the V-force of RAF Bomber Command, and that fact greatly influenced the way the programme advanced. It also influenced the specification for the V-bombers in that they were required to have bomb bays 8.8m (29ft) in length to accommodate Blue Danube (which see). The science was very well known to the scientists involved but the technology only borrowed some applications from experience with the American Manhattan Project. The total package required new designs for high explosive components, electronic apparatus for detonators, fuses and firing circuits, for the polonium initiator, the uranium tamper and the plutonium core.

A working bomb

Warhead and associated materials production is a closely guarded secret and involves no more than a handful of cabinet ministers in an ad hoc cabinet committee. There is no participation by other members of government and certainly not the remainder of either the House of Commons or the House of Lords or the general public, who are collectively embraced by the same exclusion from information and policy decisions.

Nuclear warhead research, development and implementation of committee decisions is retained within the Ministry of Defence, unlike the United States, which, implicit within the McMahon Act, saw the establishment of the Atomic Energy Commission (AEC). The AEC was abolished in 1974 and its responsibilities transferred to the Energy Research and Development Commission and the Nuclear Regulatory Commission before further merging into the Department of Energy in 1977, which has control over the US nuclear weapons programme as well as nuclear energy.

The UK Atomic Energy Authority (UKAEA) was set up in 1954, with facilities at Foulness, Orfordness and Woolwich forming the Weapons Group, which absorbed the Atomic Weapons Research Establishment (AWRE) that had previously been under the Ministry of Supply. The AWRE was moved to the Procurement Executive of the Ministry of Defence in 1973 and the name of the Weapons Group was changed to Atomic Weapons Establishment (AWE) in 1987.

The Warhead Physics Department of the AWE was broken down into four separate divisions: mathematical physics, warhead hydrodynamics, radiation physics and the explosives experimental station at Foulness. The radiation physics laboratory grew over the decades since its inception and supported two nuclear reactors (HERALD and VIPER) and a laser facility (HELEN).

HERALD was constructed as a water-moderated reactor with a consistent output of 5MW and was the most powerful of the UK's reactors of this type, providing an intense source of neutron and gamma radiation for neutron activation analysis. VIPER was unique in Europe as a pulsed fast reactor for producing intense short-duration neutron and gamma rays with peak power of 20,000MW providing data for hardening warheads.

HELEN was the UK's most powerful laser and exposed small targets to temperatures and pressures of a kind experienced in a nuclear explosion. Based on a similar device at the Lawrence Livermore Laboratory in the United States, it consisted of a very high power two-beam neodymium-glass device with an output of 10^{12}W and studied the generation of thermonuclear fusion reactions.

The warhead design department was responsible for the design of the bomb, the casing and the re-entry vehicle up to the point of manufacture. Life trials were carried out to determine safety and the condition of the stockpile as well as tests of the weapons themselves for vibration, shock and the rigours of storage.

A separate division of the warhead design department was concerned with countermeasures that an enemy may develop to inhibit the operating effectiveness of the weapons, should they be used. Tests were carried out at the US site in Nevada for X-ray, electron and laser interference as well as their survivability under conditions of an extreme electromagnetic pulse (EMP).

Test results allowed development of hardening against these effects and the fast pulse of gamma radiation from other nuclear bursts. This division also monitored foreign underground nuclear detonations. A separate division within the department developed the complex electronics required to 'safe' the weapon and to test these systems in a hard vacuum, such as would be encountered in space en route to a target.

The materials department at AWE was organised into three separate divisions for chemistry and explosives, chemical technology and metallurgy. The chemistry division monitored radiochemical analysis of fissile materials and conducted diagnostic research on the performance of nuclear devices. It also carried out corrosion studies and contributed toward the general wellbeing of physical devices as related to chemical changes likely to occur over time. An explosives technology branch provided menus for high-explosives composition, electrically initiated explosive squibs (pyrotechnic devices) and the development of long-lasting compacted charges.

The chemical and technology division was responsible for all the plastics, adhesives, rubbers, ceramics, oxides, glasses, nitride, graphite, carbon-fibre and carbon-based composites. Since 1972 it has been responsible for a wide range of new technologies employed in Chevaline and Trident projects and was located in Building A50. The metallurgy division in Building A90 carried out research on the plutonium, uranium, beryllium and various ferrous and non-ferrous materials incorporated in warhead design. It had three support departments for engineering, health and safety and a secretaries' department.

The A90 programme at Aldermaston dates to 1978 when full modernisation was recommended by radiologist Sir Edward Pochin and implemented the following year. Nine new facilities were built, including equipment for processing and handling plutonium and waste, but changes were driven by the decision to replace the Polaris SLBM with Trident and the increased capacity that this demanded. The production facility was modified to produce uranium and plutonium for the Trident II D5 warheads, which were originally to have incorporated the same number of warheads carried by the C4 Trident I.

There were many similarities between A90 at Aldermaston and the TA-55 Plutonium

BELOW Construction of the Windscale nuclear facility in Cumbria was begun in late 1947 under the Ministry of Supply as a facility for producing weapons-grade plutonium-239 utilising two air-cooled, graphite moderated reactors. (UKAEA)

ABOVE While preparations were under way to provide a deterrent with Britain's nuclear weapons capability, the UK Ministry of Defence provided – through NATO – the easternmost leg of the Ballistic Missile Early Warning System (BMEWS). Situated at Fylingdales in Yorkshire, England, it provided continuous radar surveillance to provide notice of any incoming airborne threats from the Soviet Union. *(David Baker)*

BELOW The Atomic Weapons Research Establishment had a set of facilities at Orford Ness, on a shingle spit off the village of Orford on England's Suffolk coast. First occupied as the Armament and Experimental Flight of the Royal Flying Corps in 1915, it eventually became known as the Aircraft Armament and Gunnery Experimental Establishment, hosting the AWRE where environmental testing of bomb cases and associated ancillaries took place. Pagodas such as these, still present but in a disreputable state, were used to test vibration, shock and impact such as may be occasioned during operational handling. *(David Baker)*

Processing Facility at Los Alamos National Laboratory, although the latter is exclusively dedicated to nuclear weapons research and development whereas A90 was used for the wider role of research, development and recycling of fissile material and included a waste treatment plant to strip out americium from aged nuclear warheads. Overall, the A90 modernisation programme was the biggest change at Aldermaston since its construction in the early 1950s.

Established in 1954 at the Royal Ordnance Factory, previously as an explosives filling facility, AWE Burghfield is located 8km (5 miles) south-west of Reading and the same distance from Aldermaston, designated as a prohibited place from 1976 and omitted from Ordnance Survey maps. It is the place where warheads finally come together and are integrated as a weapon system ready for deployment to relevant delivery systems, currently the Trident II missile. There are three primary complexes with several explosives bunkers on site and a large range of buildings that support final assembly in concrete igloos. It is also the place where special material, similar to Styrofoam, is produced for holding critical components of the fission core in position and for creating plasma.

AWE Cardiff is a specialised facility for fabricating and producing unique components for the nuclear weapon systems, tampers being manufactured there for the warheads from a mixture of U-238 and plutonium. It incorporated storage for depleted uranium and can produce component materials from the highly specialised hot-press, deep-draw technique. This is also used to produce pressings from beryllium powder that are used to machine the finished components and to remove contaminated beryllium from refurbished warheads.

AWE Foulness was a self-contained site situated on Foulness Island and had the great advantage of remoteness while still having good rail access. Several shock and impact tests were carried out at Foulness, with the blast tunnel being used to simulate the shock waves from a nuclear explosion. In that facility, large objects such as ground vehicles could be placed at the end of a series of stepped cylindrical sections to a maximum diameter of 10.6m (35ft). Foulness played a major part in the development of the British atomic bomb and in the production of Britain's first thermonuclear weapons. It was from here that Britain's first H-bomb was transferred to HMS *Plym*, which sailed for the Monte Bello Islands off Australia while the fissile core was flown across by Sunderland flying boat.

Over time the UK has supported ten reactors producing plutonium for its bombs, the first two being the graphite-moderated reactors at Windscale, with the next four being the gas-cooled, graphite-moderated natural uranium Magnox reactors at Calder Hall on the Sellafield site alongside the processing facility. Four more Magnox reactors were built at Chapelcross on the Solway Firth, east of Dumfries in south-west Scotland.

A serious fire at the Windscale reactor on 7 October 1957 resulted in a name change to Sellafield, to help erase the plagued name and to redirect attention toward a different future in which new safety features were incorporated and new procedures set in place. When the fire occurred the reactor was employed on the production of polonium-210 through the irradiation of bismuth-209. The polonium was used as an initiator for the British bombs.

By the early 1990s, following the collapse of Soviet communism and a change in direction of British nuclear programmes, it is believed that Sellafield had produced 3,800kg (8,380lb) of weapons-grade plutonium from these ten reactors and a further 300kg (660lb) from discharge at several civil reactors. Of this total, 100kg (220lb) would have been lost in processing, 200kg (440lb) would have been used in explosive testing, and about 1,000kg (2,205lb) was exported to the United States. About 2,800kg (6,174lb) would have been available for British weapons. This is proportionate for the production of 400 warheads, which would call for 1,600kg (3,528lb) of plutonium at 4kg (8.8lb) per bomb.

Under the 1958 Anglo-American Agreement for Cooperation on the Uses of Atomic Energy, the UK would supply the US with weapons-grade plutonium in exchange for highly enriched uranium and tritium. Also, prior to 1969 plutonium was provided by the UK's commercial reactors, despite government denials, and Sellafield was used to recover weapons-grade plutonium from nuclear fuel rods. These were then sent by road to Aldermaston in special billets for final assembly at AWE Burghfield with tritium supplied from a plant at the Chapelcross reactor site.

The achievement of the British nuclear weapons programme reached peak equilibrium in 1964 when the UK had stockpiled sufficient weapons-grade uranium and plutonium for all its weapons requirements. After that date about 8,000kg (17,640lb) of fuel and reactor-grade plutonium had been produced by the early 1990s and this was counted as part of the military stockpile. British Nuclear Fuels Limited (BNFL) played a major role in providing nuclear fuel but this work for the Ministry of Defence was no more than 10% of its overall output. When Britain joined the European Community in

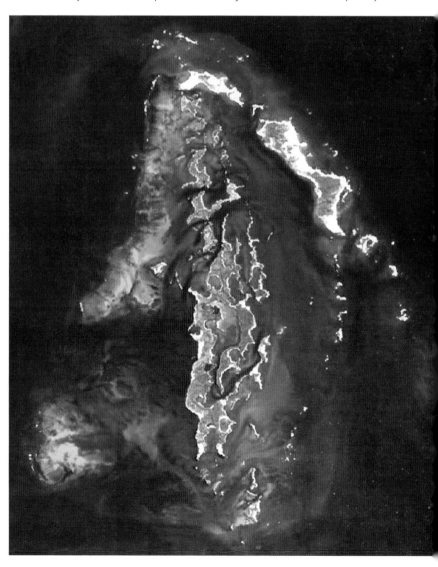

BELOW A view from space of the Monte Bello Islands where Britain's first nuclear test was to be conducted, a point of difficult negotiation with the Australian government and eventually the reason why tests with later bombs switched to the Pacific Ocean. *(NASA)*

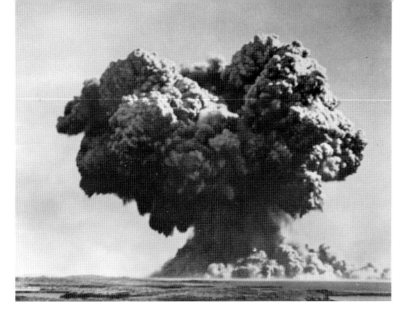

1972 there was pressure for the UK to halt the co-processing of civil and military fuel to comply with Euratom regulatory requirements. The government complied and the co-processing ended in 1986.

Britain had special requirements that it shared in Europe only with France, in that both countries have nuclear weapons and reactors for power production, the latter serving not only domestic power requirements but also nuclear-powered submarines as part of their national defence infrastructure. Moreover, there has been a gradual shift over the decades away from close cooperation with exchange agreements involving the United States and more towards an independent stance on the production of weapons fuel.

Britain's somewhat limited nuclear stockpile requires relatively little highly-enriched uranium (HEU) to the required 90% and from the early 1950s it was decided to achieve this at a plant at Capenhurst, Cheshire. It started operation in 1953 and began producing significant amounts of HEU four years later. From 1958 on the US provided HEU in exchange for plutonium, as discussed above, but from the 1980s the UK procured it from America on a purchase basis. Of the total produced at Capenhurst, which was about 6,000kg (13,230lb) added to 7,000kg (15,435lb) from the US, 2,700kg (5,950lb) has been used for British nuclear-powered submarines and 10,000kg (22,050lb) was available for warheads or stockpiling.

In the early 1980s the British switched from enrichment by gaseous diffusion to the more efficient gas centrifuge technique, making the process more cost-effective and saving

costs on procurement from the US. But this also began to make the UK significantly more independent on the production of appropriate fuels and on the ability of the UK to maintain a supply should political changes affect the supply from America. The British government saw this as an insurance policy against retraction on the part of the Americans. This decision also affected the supply of tritium.

Until the mid-1970s the UK bought its tritium stocks directly from the United States before switching to a domestic production capability. This was to preserve an independent supply less reliant on Anglo-American cooperation, and also to save money and create additional jobs in the British nuclear industry. But it was also as a hedge against US concerns about nuclear proliferation and signs of an impending set of restrictions from the American government on foreign exports, even exclusive deals done with a close ally.

Tritium could be provided from the same British reactors that produced plutonium, by irradiating lithium-6 rather than uranium-238. Natural lithium contains 7.42% lithium-6 and 92.58% lithium-7. BNFL built a tritium production plant at Chapelcross for this job that became operational in 1980. Thus did Britain become only the fifth country to have a domestic tritium production capability; in addition to the USA, Russia, France and China, its place at the metaphorical 'Top Table' was established.

Testing the bomb

Three decades earlier, in September 1950 the British Chiefs of Staff recommended to Prime Minister Attlee that the bomb, when it was ready, should be tested on the Monte Bello Islands off the north-west coast of Australia. Several other locations within the British Commonwealth had been considered, including several sites in Canada, while there had been faint hope of getting a deal with the United States for use of its Pacific test sites. Having received a recommendation for Australia as the test site for the atomic bombs, Attlee sent a letter to Australian Premier Robert Menzies about the proposal on 16 September 1950, followed by a survey party from the UK visiting the Monte Bello site.

In March 1951 Attlee sent a formal request to use the site for tests scheduled to begin in October 1952, which was granted after the Australian elections returned Menzies to power in May 1951. Probing requests to the United States had been sent out in August 1950 but use of the proposed Eniwetok site was denied by the US Joint Chiefs, citing the McMahon Act as prohibiting cooperation with other states in the development of atomic weapons.

After persistent British attempts to reverse this, Secretary of State Dean Acheson offered a site in Nevada but only on the basis that the UK allowed the Americans full access to all the bomb's technical secrets and gave them access to inspection of all the components by the Atomic Energy Commission. The British refused, but a concession was offered by the Americans in which they would be allowed sufficient access to the design details only to reassure them that its explosive yield would not exceed 25KT. Again, the British refused.

But Penney was insistent that the British government take up the American offer, seeing in it the bridge to build back a consolidated relationship on the development of atomic weapons, and he flew to the US several times to discuss the matter. The Americans felt progress was being made and integrated the British bomb into their forward planning for the impending Upshot test series. This would have required about 50 British personnel to fly out and would cost the UK about $1 million. After the British elections that returned Churchill as Prime Minister, all question of acquiescence to

American terms evaporated. A joint declaration confirming the decision to use Monte Bello was issued on 19 February 1952.

The first test of a British A-bomb got under way when the initial assembly left Foulness at 8:30am on Thursday 5 June 1952 to be loaded aboard the frigate HMS *Plym* moored at Stangate Creek, Shoeburyness. The following day the second assembly was loaded and the vessel set sail on an eight-week voyage around the Cape of Good Hope and across the southern Indian Ocean bound for Monte Bello, where it arrived on 8 August. Escorted all the way, the plutonium core left by Handley Page Hastings from RAF Lyneham, Wiltshire, and flew to Singapore via Cyprus, Sharjah and Ceylon (now Sri Lanka), where it was loaded aboard a Sunderland flying boat for the rest of the trip to Monte Bello, arriving on 18 September.

For the test, named *Hurricane*, the bomb was encased in a watertight caisson suspended 27.5m (90ft) below HMS *Plym*, which was moored in a lagoon off the western shore of Trimouille Island. It was successfully detonated on 3 October delivering a yield of 25KT, making Britain the third country to detonate a nuclear device. Ironically, the fear that the Russians might smuggle an atomic weapon into a British harbour had inspired Penney to configure the test in this way to evaluate the effects of such a detonation on the hull of a ship!

Post-detonation evaluation revealed that the bomb was 20% more efficient than the first American bomb of this type, which had been detonated in mid-1946, and that it was a great

BELOW Vickers Valiant XD818 was used in drop-tests at Orford Ness, England, and on 15 May 1957 for releasing Britain's first 'layer-cake' weapon contained within a Blue Danube casing. *(Chris Gibson)*

RIGHT The Short Granite shot in Operation Grapple produces a successful detonation on 15 May 1957, high in the atmosphere to minimise fallout but delivering a disappointingly low yield of 300KT. *(Via David Baker)*

BELOW Retention rings for containment within the bomb casing adapted from the Blue Danube device look like cartwheels, prior to the Orange Herald detonation on 31 May 1957, not a true thermonuclear device as proclaimed at the time but a fusion-boosted fission weapon which delivered a yield of 720KT. *(David Baker)*

improvement. It had cost the British government £160 million to get a working bomb, about £4 billion in 2017 money, which was about one-quarter of what the Americans spent to get to the Trinity Test in 1945. Further work to refine the design and to create a safe and reliable weapon continued apace and that prompted planning for additional tests.

Between 1952 and 1957 the UK conducted a total of 12 tests in Australia: three at the Monte Bello site, two at Emu Field in South Australia and seven at Maralinga close by and a part of the Woomera range where rockets and missiles were being tested. Penney planned a wide range of tests and they would commence with the *Totem* series to decide the composition of plutonium to be employed for the first operational bomb, Blue Danube, or Mk 1, which was to be carried by the V-force.

This information was required to plan the doubling of plutonium production so that 200 bombs could be deployed by 1957. Specifically, it was to determine the amounts of the isotope Pu-240 would have on the yield of a nuclear weapon. It was decided to add two new Magnox reactors at Calder Hall, which would produce bomb-grade plutonium. These tests would help decide the optimum burn-up rate for the dual-purpose Magnox reactors. Reactors optimised for power production use higher burn-up rates and produce plutonium with a higher Pu-240 isotope composition compared to reactors producing weapons-grade plutonium.

Totem 1 took place on 14 October 1953 and delivered a yield of 10KT, about twice that anticipated, followed by *Totem 2* on 26 October, when a yield of 8KT was obtained, almost three times the expected level. As a sign of rapidly converging interest from the United States, two Boeing B-29 Superfortress bombers were allowed to fly out of Richmond, New South Wales, and sample the air from a distance no closer than 650km (400 miles) downwind of Emu Field. The US would reciprocate by allowing two RAF Canberra bombers to sample thermonuclear fallout from the Castle tests in the Pacific Ocean during early 1954.

Tasked with developing an efficient stockpile of Blue Danube bombs and to make efficient use of the costly fissile materials, Penney also had to manage the start of research into the UK's own thermonuclear (hydrogen) bomb, which also had to be carried to completion. Details of the British H-bomb's origin are contested, but it is known that on 16 June 1954 a small cabinet sub-committee, chaired by Churchill, decided to develop the British 'Super'. In the spring of 1955 advertisements

appeared for theoretical physicists to work at Aldermaston and at least one scientist who joined the team spoke of his assignment to working on deciphering the Ulam-Teller solution to providing a thermonuclear device based on fusion of low energy nuclei.

Meanwhile, between 27 September and 22 October 1956 four *Buffalo* tests were conducted at the Maralinga site. The first and last were tower tests of the new Red Beard (see below), while the second, on 4 October, was the ground test of a low-yield Blue Danube bomb of 1.5KT and the third, on 11 October, was an air-dropped Blue Danube. Carried aloft by a Vickers Valiant (WZ366) of No 49 Squadron to an altitude of 9,140m (30,000ft), the Blue Danube for this first air-drop was descaled to a yield of 3KT to limit radioactive fallout and detonated at an altitude of 152m (500ft), rather than expose the region to the device's designed value of 40KT exploded at 366m (1,200ft).

The Red Beard device had been produced for use as the first British tactical nuclear weapon, emerging at a time when the United States was rapidly expanding the development, and deployment, of such weapons. The first one had an expected yield of 16KT from a mixed core of weapons-grade plutonium and weapons-grade uranium-235, but the Mk 1 bomb for operational deployment was designed for a yield of 15KT, while the Mk 2 had a yield of 25KT, the latter available in two versions for

high-altitude or low-level toss-bombing. It was designed to be carried on the RAF's Canberra light bomber as well as the V-bombers and by the Blackburn Buccaneer, Sea Vixen and Scimitar of the Fleet Air Arm.

The *Buffalo* tests verified that, quite independently, the British had by 1956 acquired a light tactical nuclear warhead, a

ABOVE Grapple C produced a 1.8MT detonation on 8 November 1957, Britain's first true thermonuclear device, released from a Vickers Valiant over Christmas Island.
(Via David Baker)

LEFT An aged and grainy photograph of some of the mostly unnamed Blue Danube development team responsible for delivering Britain's first production nuclear weapon.
(Via David Baker)

ABOVE **The physics package for the Blue Danube device, superficially similar to the bombs used for the weapons tests over Christmas Island but here showing the detonators spaced evenly around the spherical bomb.** *(Via David Baker)*

high-yield weapon for V-bombers and a very high-yield weapon for use in the Blue Streak ballistic missile, to which the UK had made a commitment the previous year. The tests had provided valuable data on the effects of nuclear weapons on a wide range of military vehicles and redundant aircraft placed strategically, and on the physiology of 253 servicemen, known as the Indoctrinee Group.

These men, including 70 Australian officers from its three services and five from New Zealand, were placed as close as possible to two *Buffalo* detonations. In the first they were situated in trenches 7.24km (4.50 miles) from the bomb. In the second – the low-yield Blue Danube ground detonation for measuring shock and seismic effects – 65 officers were positioned 2.74km (1.7 miles) from ground zero in a viewing stand while 24 other men were positioned in slit trenches much closer and four were placed in a Centurion tank.

The third cluster of tests, named *Antler*, tested low-yield devices for use in surface-to-air guided weapons, in the Red Beard tactical bomb and to advance the technology of using small fission bombs to trigger megaton-range weapons. They began with a 0.93KT detonation on 14 September 1957, continued with a 5.7KT detonation on 25 September and concluded on 9 October with a 26.6KT explosion, bringing to an end the main sequence of Maralinga tests.

However, between 1958 and 1963 some 550 experiments were conducted at the site

on separate tests designed to carry out a very wide range of investigations contributing to the remarkable series of developments, the vast majority of which remain highly classified, that have propelled the UK to the forefront of nuclear weapons design and to the perfection of both fission and fusion weapons across a very wide yield range. It would be wrong to think that all these tests were contributing to the bomb projects, for while that was their primary justification and purpose the lessons learned were also applied to the efficiency and safety of nuclear power plants across the UK.

As identified elsewhere in the section describing the development of US thermonuclear weapons, the Russian nuclear test in August 1949 stirred minds in America and Britain regarding the availability of technology for the H-bomb and, to a lesser extent, their moral position regarding the desirability of building such a bomb. There was little doubt that the Russians would seek to acquire an H-bomb at the earliest opportunity and that the early development of British and American bombs of this type was both desirable and prudent.

It was to be a year after the US decision to build an H-bomb, taken by President Truman on 31 January 1950, before the Ulan-Teller solution had been presented to make such a bomb possible. The successful US detonation of a thermonuclear device on 31 October 1952 had verified the calculations and on 16 June 1954 a small cabinet committee agreed to go ahead with a British H-bomb. Churchill informed President Eisenhower of this decision on 26 June during a visit to America and informed the Canadian Prime Minister three days later.

It is interesting to note that upon his return to the UK, Churchill was met with a general view that this could serve as a restraint on American 'adventurism', as there appeared to be little concern about Russian 'aggression'. The public announcement of the H-bomb decision was issued on 17 February 1955. Not only did this decision place a great deal of strain on the capacity of Britain's nuclear weapons industry; implicit in the decision was also a commitment to develop and deploy a strategic ballistic missile with a range of

3,200km (2,000 miles) from launch pads in the UK, a missile against which, at the time, there was no known defence.

This represented a major expansion of UK nuclear deterrent capability, of high-yield weapons development and of unique delivery systems. It came when the United States was expanding its own nuclear deterrent with the commitment to Atlas and Titan intercontinental ballistic missiles (ICBMs) and was willing to share rocket technology with the UK in order for Britain to build what would be known as the Blue Streak medium-range ballistic missile (MRBM). By the end of 1955 the plan was for a deployed field of Blue Streak MRBMs, in addition to strategic and tactical nuclear bombs delivered by air.

The design of a British H-bomb was completed between 1954, when its scientists were able to obtain data from the US Castle test sequence in which a variety of configurations were evaluated, and 1956. This provided scientists and engineers with direct access to American research which, when integrated with that orchestrated by William Penney, produced a viable design concept for the British bomb. An integrated sequence of tests in Australia had helped push forward the technology the British still needed for their thermonuclear weapon, while developing the greater efficiency of the fission devices, Red Beard and Blue Danube.

There was a level of difficulty about the two tests to be conducted in support for the British H-bomb. The Australian government had been given assurances that these tests would not involve yield levels more than 2.5 times that of the *Hurricane* test. There were already concerns about the level of radioactive contamination from fallout, and the Castle tests in 1954 had begun to generate concerns around the world. The landlocked site at Maralinga was not the best place to release unexpected qualities of fusion-induced fallout.

BELOW The layout for the Blue Danube bomb case, the largest and most unwieldy of Britain's nuclear weapons, together with dimensions and centre of mass.
(Chris Gibson)

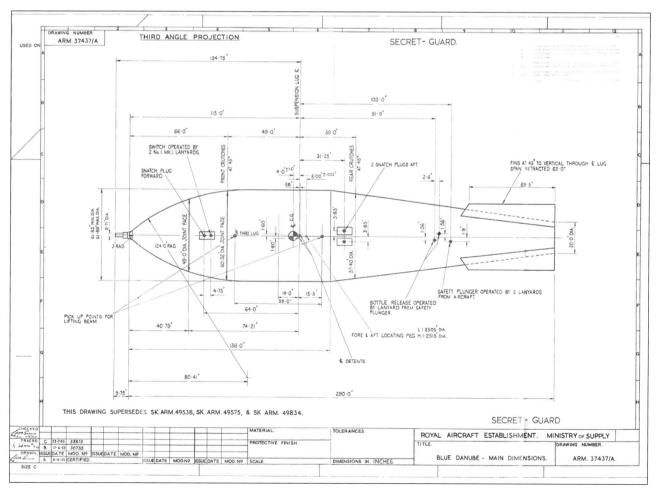

Known as *Mosaic 1* and *Mosaic 2*, these special tests, crucial to the definitive thermonuclear weapon, were conducted in 1956 between the *Totem* and *Buffalo* atomic bomb related tests. The first, conducted on 16 May, delivered a yield of about 15KT but the second, on 19 June, delivered 98KT, a figure which was four times that of the *Hurricane* test. That information was withheld from the Australians for three decades. Nevertheless, the tests provided the required data and the British moved to fully exploit their renewed level of cooperation with the Americans.

The two *Mosaic* tests were a vital part of preparation for the thermonuclear bomb test and helped fill the gaps in scientists' understanding of the exact action of the high-energy neutrons released in a fusion device. But the British scientists had at best an uncertain knowledge of the Ulam-Teller principle and adopted an independent approach for the initial series of *Grapple* tests, which began on 15 May 1957 with the test of *Short Granite*, a device built around the same logical principle applied by the Russians with Joe 4, a layer-cake concept whereby a layer of lithium deuteride surrounds the fissile core which is in turn surrounded by a layer of U-238.

The device was dropped from Valiant bomber XD818 of No 49 Squadron, with recording gear, at Malden Islands, south of Christmas Island, where the bomb was released. Final arming was by the crew on board the aircraft. With a total weight of 4,500kg (10,000lb) in its Blue Danube case, the bomb was detonated at an altitude of 730m (2,400ft) some 52 seconds after release, exploding with a yield of no more than 150KT. The Christmas Island site had been chosen in late 1955 because of its isolation and relatively unpopulated area.

The second test on 31 May was dubbed *Orange Herald* and constituted a 'fall-back' fission design in the event that testing of the full thermonuclear fusion device was halted by a global ban on atmospheric testing, of which there was a distinct possibility due to public outcry over radiation levels resulting from the numerous tests already conducted by America and Russia. Detonated at a height of 700m (2,300ft) after release from Valiant XD822, it had a yield of 720KT. The third in this initial series, named *Purple Granite*, was detonated on 19 June from Valiant XD823 and ran to 100KT.

There was disappointment in that the devices had not performed as well as expected. Clearly there was still work to be done by

RIGHT The interior layout of Violet Club, the only wholly British bomb deployed as a stopgap until a true thermonuclear device could be deployed. Essentially the Green Grass bomb tested in 1957, it was oversold as a thermonuclear device but was made to impress the Americans and expand cooperation, which it did the following year.
(Brian Burnell)

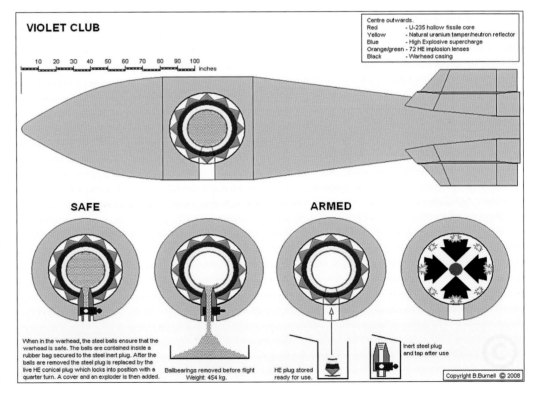

VIOLET CLUB

Centre outwards.
Red - U-235 hollow fissile core
Yellow - Natural uranium tamper/neutron reflector
Blue - High Explosive supercharge
Orange/green - 72 HE implosion lenses
Black - Warhead casing

10 20 30 40 50 60 70 80 90 100
inches

SAFE

ARMED

When in the warhead, the steel balls ensure that the warhead is safe. The balls are contained inside a rubber bag secured to the steel inert plug. After the balls are removed the steel plug is replaced by the live HE conical plug which locks into position with a quarter turn. A cover and an exploder is then added.

Ballbearings removed before flight
Weight: 454 kg.

HE plug stored ready for use.

Inert steel plug and tap after use

Copyright B.Burnell © 2008

scientists at Aldermaston and elsewhere to prepare for a true thermonuclear reaction. After the *Antler* atom bomb tests in September and October 1957, the thermonuclear development trials resumed on 8 November 1957 with *Grapple X/Round C*, which was released from Valiant XD824 at an altitude of 685m (2,250ft), delivering a yield of 300KT – much higher than expected. This was one of Penney's 'fall-back' tests and provided further verification of the boosted fission design.

The first full thermonuclear test, designated *Grapple Y*, took place on 28 April 1958 when Valiant XD825 dropped a 2MT bomb which detonated at 2,370m (7,700ft), the first unequivocal British-designed and built, two-stage fusion bomb of the Teller-Ulam type. It was followed on 22 August by the *Grapple Z/Pendant 2* test in the 26-42KT range, detonated from a height of 450m (1,476ft) suspended beneath four balloons. The test was of a fission device designed to trigger a thermonuclear fusion reaction.

The next test (*Grapple Z/Flagpole 1*), on 2 September 1958 was an air-drop bomb detonated at an altitude of 2,850m (9,140ft) after release from Valiant XD822, delivering a yield of almost 3MT. This was followed nine days later by a drop from Valiant XD827 of a similarly sized device (*Grapple Z/Halliard 1*). The last British atmospheric test took place on 23 September (*Grapple Z/Burgee 2*), essentially a repeat of *Grapple Z/Pendant 2*, the 21st British nuclear weapons test. There had been plans for another test, *Grapple Mike*, but the nuclear test moratorium introduced by the Russians from 31 March, and by the Americans later that year, stopped that happening.

The British agreed to halt testing contingent on the Americans providing the British with nuclear weapons information through an amendment to the McMahon Act. In return, the Americans got the use of Christmas Island. Further to that, when John F. Kennedy became President in January 1961 he pursued a policy supporting a comprehensive test ban treaty, but that was not achieved and from November the moratorium had been lifted and atmospheric tests resumed. By this date the British had acquired the thermonuclear bomb and tactical atomic weapons and were already acquiring

a wide range of American nuclear shells and rockets for the army, as defined elsewhere in this book.

The more inclusive arrangement whereby the British and the Americans now shared nuclear design and weapons test information extended to participation in the US underground nuclear test series at the Nevada site. It followed a decision between President Eisenhower and British premier Harold Macmillan at a meeting in March 1957 for the Americans to supply the RAF with 60 Thor intermediate-range ballistic missiles (IRBMs) for deployment in the UK.

After the launch of Sputnik 1 on 4 October 1957, and the de facto existence of a Soviet ICBM, a further meeting on 23–25 October sought much closer Anglo-American cooperation. On 2 July 1958 Eisenhower signed the amendments to the McMahon Act and a new era of cooperation ensued. It was a surprise to the Americans to learn that British ingenuity and technology development had created certain fission and fusion weapons that were in advance of their own, and written allusion to this is prevalent in the documents exchanged between the international partners. Never had the Americans divulged as much about their programmes as they did following initial technical meetings beginning on 25 August 1958.

The Americans gained as well in the unique configuration of the British 'peanut' design of the geometry of the two stages, rather than the dual spherical shape used by the Americans. But the British acquired technology information about the way uranium in higher burn-up reactors could produce useable plutonium for bombs, knowledge which would help with the civilian reactors. Cooperation also extended to sharing information between British intelligence, the US Central Intelligence Agency (CIA) and the US Defense Intelligence Agency (DIA) about Russian nuclear tests.

As cooperation evolved, approval was given for the British to acquire the Skybolt stand-off missile to replace the Blue Steel weapon carried by Victor and Vulcan aircraft of Bomber Command, and that story is told in a later section. When Skybolt was cancelled, agreement was reached between Kennedy and Macmillan in December 1962 for the British to acquire Polaris submarine-launched ballistic missiles (SLBMs) for British-built submarines as the Royal Navy replaced the Royal Air Force as carrier of the country's strategic nuclear deterrent.

When the test moratorium ended in September 1961 the British had not resumed testing at Maralinga or Christmas Island. Instead, testing would take place at the US test sites, beginning on 1 March 1962 and continuing until 26 November 1991, during which period the British conducted 24 tests, all of which were in the kiloton range in support of the British-designed and built warheads for the Polaris and Trident missiles.

The exchange of information resulting from the amended McMahon Act of 1958 provided valuable assistance to both the British and American programmes on a reciprocal basis, a level of cooperation which forced closer ties within the respective national security agencies as converged data about Soviet nuclear tests and weapons deployment became of paramount importance during the closing decades of the Cold War.

In a broader context, weapons effects research proved highly relevant to the expanding role of governments in protecting the civilian population in the event of a hostile nuclear exchange, and agreements were drawn up between the US, the UK, Canada, Australia and New Zealand to apply those lessons. Research into weapons hardening advanced considerably. If there was ever a defining measure of the 'special relationship' between the US and the UK it is epitomised by the extraordinary levels of cooperation forged first between Eisenhower and Macmillan in 1958 and then between Kennedy and Macmillan in 1962.

British tests

The UK conducted 88 tests between 3 October 1952 and 26 November 1991, with a similar number of devices fired, all but 24 being at the Monte Bello Islands, Emu Field and Maralinga, Australia, and Kiritimati and Malden Island in Kiribati. The remainder were conducted at the Nevada Test Site in the United States. With a declared cumulative yield of 9.282MT, the total UK contribution to the global nuclear test log is 1.7%. One of the reasons why there were so few British tests was the expense involved and the availability of unprecedented access to American data from similar devices.

The British stockpile

Cooperation with the United States saved the UK a considerable amount of money but, until the amendment to the McMahon Act, Britain made plans for a V-force equipped with the Blue Danube atomic bomb of similar yield to the bomb dropped on Hiroshima. Subsequently, and with the decision to provide a warhead for the Blue Streak MRBM, there was a need to develop a high-yield thermonuclear device.

Against a background of shifting policy and procurement decisions, the work of the nuclear weapons team and its comprehensive network of facilities was tasked with an ever-changing set of requirements. The independent period up to 1958 was the first phase; the period after, which involved US cooperation, was the second. During the first phase the British achieved a refined fission bomb, a multistage fusion bomb and a 'layer-cake' design. It placed Blue Danube in production and some Yellow Sun thermonuclear bombs. The second phase began a shift toward the use of American designs in UK-manufactured bombs and the Red Beard tactical bomb, and that reliance on US designs drastically reduced the requirement for tests.

In brief, delivery systems relied initially on the V-force, which became operationally active with the Valiant and Blue Danube in 1957. The expanded plan was for the Blue Streak ICBM to be ready and operational by 1965 but delays prevented that before it was cancelled in 1960. With non-storable propellants, as a weapon it was considered too vulnerable to survive a pre-emptive attack. In the interim, in 1957 the Americans agreed to deploy 60 Thor IRBMs to Britain until Blue Streak was ready. The first Thor was deployed in 1958 but they were withdrawn by the end of 1963.

Threatened with redundancy due to enhanced Soviet air defences, the V-force was considered to require a stand-off weapon for penetrating enemy airspace and in 1954 the Blue Steel programme was initiated. Blue Steel became fully operational in early 1963 but the system was retired by 1970. Both Thor and Blue Steel carried nuclear devices originating in American systems – in fact Thor carried the Americans' W49 warhead in Mk 2 re-entry vehicles, although the British were given the option of replacing this with a UK warhead. They never did.

During the development of Blue Steel, under the new 1958 accord, the Americans offered the proposed Skybolt air-to-surface missile, essentially a stand-off bomb, but with a much greater range than its British equivalent. Billed as a successor to Blue Steel, it would have been carried under the wings of a Vulcan bomber, much as it was planned to be carried under the wings of a US Air Force B-52. The Kennedy administration cancelled Skybolt at the end of 1962 on cost grounds, by which time they had offered Britain the Polaris SLBM. That went through, with the Navy taking over the strategic nuclear deterrent from the RAF in 1969 and continuing to do so to this day with Trident missiles.

Thus was the tortuous path to Britain's

LEFT The clockwork fuse for the Blue Danube bomb. *(Via David Baker)*

CENTRE To ensure symmetrical firing of the high explosive initiators, the 32 devices had 64 firing wires connected in parallel, achieving a higher level of certainty that they would fire. *(Via David Baker)*

BELOW To ensure maximum security, two manufacturers were placed under contract to produce elements of the Blue Danube ground proximity fuse. *(Via David Baker)*

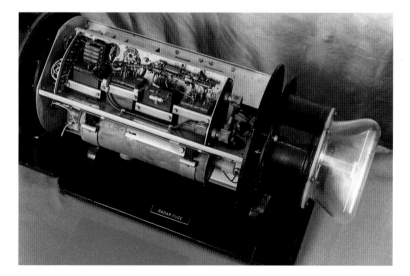

nuclear stockpile laid down through a series of changes and policy directives that typically failed to outlive the precursor development period. The following description of individual weapons identifies the land, sea and air systems developed and/or operated by the UK, including the prolific array of American nuclear warheads and weapons operated by the British during the Cold War.

There was one very big advantage in closely aligning British national interests with the defence establishment in the United States and, through them, acquisition of the American submarine-launched Polaris and Trident missiles. It locked the United States irrevocably into the defence of the United Kingdom and, de facto, into that of Europe itself. Because if Russian radars saw a missile ascending through the waters of the Atlantic Ocean, the North Sea or the Indian Ocean, they would not know whether it was British or American, and, because of that, total Anglo-American alliance would be involved from the outset. It was a chain that shackled the US to the affairs of Europe.

Blue Danube

Otherwise known as the Mk 1 and built to Operational Requirement OR 1001, issued on 9 August 1946, the first British atom bomb was delivered to RAF Wittering in November 1953, with No 1321 Flight flying Vickers Valiant bombers operational from April 1954. However, integration trials with aircraft and bomb did not begin until mid-1955 with tests off dummy cases at the Orfordness firing range, the first taking place on 6 July. The RAF carried out its first assembly and dismantling of live radioactive components on 27 July. During that year RAF stations at Gaydon and Wyton joined Wittering as the foundation of the V-force within No 3 Group Bomber Command.

Blue Danube was rare among weapon systems in that it was built and deployed before testing, and modifications introduced along the way provided improvements, greater efficiency in handling and operation and provided an evolving series of what were, in essence, hand-made devices. The biggest challenges of all had been in theoretical modelling of the device and in the electronic circuitry, without which it would not have worked. This pushed along an entire industry, which had vast – but totally unattributed – benefits.

The design was based initially on the package developed for the *Hurricane* test, which was itself a conceptual parity with the American 'Fat Man' bomb but differed in that it incorporated a light and internally braced case with an explosive sphere constructed from a combination of 12 pentagonal and 20 hexagonal shapes. The first bombs had a plutonium core and one was tested with a pure uranium core, but most bombs assembled for the RAF had a plutonium/uranium-235 core that had definite advantages in reliability and predictability.

The design of the bomb was broken down into six separate packages, comprising the outer ballistic casing, suspension for the central physics package, the fuses, firing mechanism and detonators, the high-explosive shell

BELOW The ground plan for the nuclear storage facility at RAF Barnham. *(Via David Baker)*

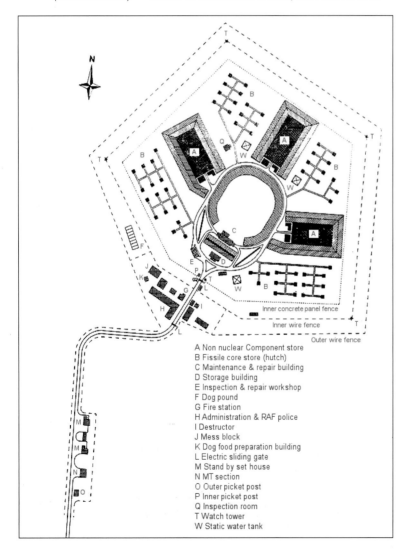

A Non nuclear Component store
B Fissile core store (hutch)
C Maintenance & repair building
D Storage building
E Inspection & repair workshop
F Dog pound
G Fire station
H Administration & RAF police
I Destructor
J Mess block
K Dog food preparation building
L Electric sliding gate
M Stand by set house
N MT section
O Outer picket post
P Inner picket post
Q Inspection room
T Watch tower
W Static water tank

Inner concrete panel fence
Inner wire fence
Outer wire fence

and tamper, core and urchin (a tiny amount of polonium 210 to ensure initiation of the reaction), and the handling fixtures and fittings. The casing and the fixtures were designed at the Royal Aircraft Establishment, Farnborough, an introduction to nuclear weapons that was to increase and diversify during the thermonuclear bomb development period and the design of the WE177. This work was carried out clandestinely at Farnborough's Airfield Radio Laboratory (ARL).

Somewhat less furtively, the main bomb case was fabricated by Hudswell Clarke of Leeds, Yorkshire, a company which had made its reputation manufacturing railway locomotives and would move on to rocket production. But the secrecy that surrounded every other conceivable element of the bomb was undone in grand style by the placing of a neat, brass plaque on the side of Blue Danube proudly revealing the serial number and the customer's name – Atomic Weapons Research Establishment. No worker was in any doubt about the nature of his job!

Blue Danube had a total length of 7.37m (24.17ft) and a maximum diameter of 1.57m (5.16ft) supporting four pop-out tail fins, the whole assembly having the unfortunate characteristic of flying along with the aircraft after release! Blue Danube weighed about 4,636kg (10,222lb), which had defined the requirement for the bomb bay size on the Valiant, the Victor and the Vulcan. The bomb had been designed to provide an explosive yield of 10–12KT, similar to that of the Hiroshima bomb, and to be readily operated by RAF personnel under a variety of operational circumstances.

Unlike several American bombs Blue Danube could not be armed in flight, but four separate triggers were incorporated including radar detonation, barostat, timer and inertia switch. On one alarming test run with a dummy package the bomb hung up inside the Valiant. After the aircraft returned with the bomb the ground crew released the doors, causing the bomb to drop to the ground – a shock that would certainly have detonated the device had it been live!

The original idea was to deploy about 200 bombs, for which an anticipated 800 were planned, but actual production and stockpiles fell considerably short of that figure, no more

than 58 being produced with a maximum 20 on full alert potential. The bomb was 'operational', in a qualified sense, between 1953 and 1962. The bombs were stored in special facilities at RAF Barnham, Suffolk, and RAF Faldingworth, Lincolnshire, which would have provided Scampton, Finningley and Coningsby with weapons in the event of a high alert level.

Despite intentions to the contrary, Blue Danube was not as effective as an operational weapon as had been hoped, due to the experimental nature of the programme and technical limitations, including lead-acid accumulators and the radar altimeters. Modifications were made which improved the effectiveness of Blue Danube as a reliable weapon but it could never be considered a definitive answer to the need for an effective deterrent.

ABOVE Storage igloos at Barnham for housing plutonium cores, elements of which survive today, albeit in very poor states of repair. *(Via David Baker)*

BELOW A technician works on the Blue Danube bomb case prior to testing, the image revealing ground-handling cradle pick-off points and hoist attachments for the 4.5-tonne device *(Chris Gibson)*

unboosted fission device using a composite core of weapons-grade plutonium and U-235, effectively a safer device and one minimising the use of costly fissile materials. It was essentially a second generation Blue Danube but with selective yields of 15KT for the Mk 1 version and 25KT for the Mk 2. The Mk 2 was available in two versions, for high-altitude and low-altitude delivery respectively. It used the same ballistic shape for the casing and had a length of 3.66m (12ft) incorporating pop-out fins, a maximum diameter of 0.71m (28in) and an unballasted weight of 794kg (1,750lb).

Nevertheless, Blue Danube was a British bomb built by British scientists and fabricated by British engineers, at a time when similar British brains had accelerated the development of the American bomb in the Manhattan Project, and gives the lie to those who claim it all came from the United States. While being the product of varied opinions on the morality or otherwise of nuclear weapons, Blue Danube's conception is in no doubt at all, its father being a very British bulldog!

Red Beard benefited from improvements to the Blue Danube as the two weapons were very similar in several respects, not least modification to the electrical circuits and the fusing systems. Twin ram-air turbines were located in the nose, providing electrical power after release of the bomb from the carrier-plane. This avoided the unpleasant prospect of pre-release detonation through stray discharges that were thought to be a potential risk with Blue Danube, which had six lead-acid batteries that had to be quickly installed before take-off.

Red Beard

Designed to OR 1127, Red Beard was the UK's first tactical nuclear weapon and was an

Red Beard was late in entering the inventory as a tactical weapon, being introduced

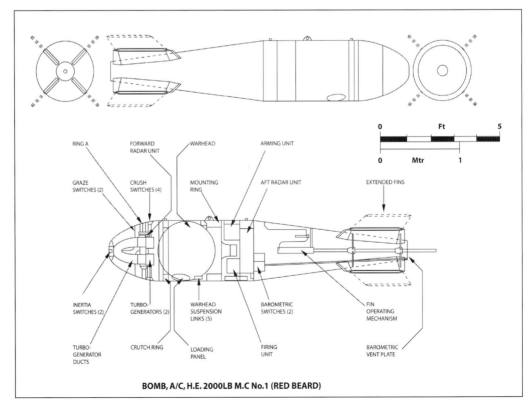

RING A
FORWARD RADAR UNIT
WARHEAD
ARMING UNIT

GRAZE SWITCHES (2)
CRUSH SWITCHES (4)
MOUNTING RING
AFT RADAR UNIT
EXTENDED FINS

INERTIA SWITCHES (2)
TURBO-GENERATORS (2)
WARHEAD SUSPENSION LINKS (5)
BAROMETRIC SWITCHES (2)
FIN OPERATING MECHANISM

TURBO-GENERATOR DUCTS
CRUTCH RING
LOADING PANEL
FIRING UNIT
BAROMETRIC VENT PLATE

BOMB, A/C, H.E. 2000LB M.C No.1 (RED BEARD)

into service in 1961, several years after the Americans had introduced tactical nuclear weapons to their UK-based tactical fighter-bomber units in East Anglia. The ability to produce smaller and lighter fission devices broadened the application of nuclear weapons and these were matched with Scimitar and Buccaneer types for the Royal Navy and the Canberra, Valiant, Vulcan and Victor for the RAF. Initially Red Beard had been considered for the 'thin-wing' Javelin and the English Electric Lightning. Evaluation had also been made of qualifying the Sea Vixen for Red Beard but they were never deployed with that capability.

The move to equip tactical aircraft with low-yield fission bombs nevertheless required special techniques for deploying the bomb to its target and the RAF and the Royal Navy used a variation of the American toss-bombing technique where fighter-bombers would pull up in the first half of a loop, release the bomb close to the vertical and level-up at the top to dash away from the target in the opposite direction as fast as possible. Invariably, the pre-programmed arming and detonation sequence would have the bomb detonate at higher altitude than the returning aircraft.

The RAF deployed a maximum 80 bombs of this type between 1963 and 1965 before reducing the inventory to 40 during 1968 prior to retirement by the end of 1969. The Royal Navy introduced Red Beard in 1962 and ran up to an inventory of 30 between 1964 and its retirement in 1970. Of the total maximum stockpile of 130, 48 were held at RAF Tengah, Singapore, to support the UK's commitment to the South East Asia Treaty Organisation (SEATO) with an additional 48 in Cyprus supporting the Central Treaty Organisation (CENTO).

Yellow Sun

The desire for a very large bomb led to Britain's first thermonuclear device via a failed project known as Violet Club, taking its name from the Rainbow Codes first used during the Second World War to hide secret weapons or captured devices. As such the Mk 1 Yellow Sun was the last wholly British-built bomb and was succeeded by Yellow Sun Mk 2. Both were held within a case 6.4m (21ft) in length and 1.2m (48in) in diameter. The Mk 1 weighed 3,290kg (7,250lb) and the Mk 2 770kg (1,700lb). Neither employed parachute retardation after release but relied on a blunt nose to achieve maximum drag

ABOVE A Red Beard round at RAF Cosford, with folded pop-out fins clearly visible. (Chris Gibson)

BELOW The general layout of the UK's thermonuclear device, which followed the essential design guidelines of the Teller-Ulam concept. (Via David Baker)

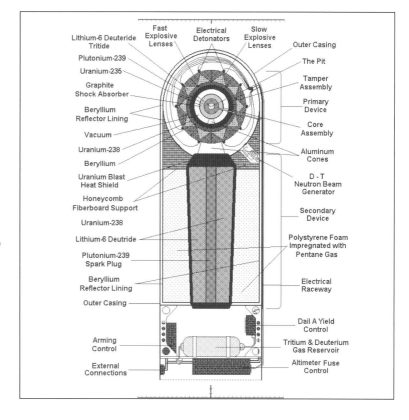

ABOVE The Yellow Sun case was designed to carry a powerful thermonuclear bomb. To facilitate a higher induced drag it had a bluff nose which prevented the device reaching transonic speeds, which had been shown to adversely affect the barometric fuse on Blue Danube. *(Via David Baker)*

BELOW Yellow Sun Mk 1 and its Green Grass warhead seen here as the cylindrical centre section, a case also applicable to Red Snow, in service for a decade and at the time the mainstay of the nuclear deterrent. *(Chris Gibson)*

BOTTOM Red Snow installed in the centre section of Yellow Sun Mk 2, which was developed from the American W28 warhead. Ballast was necessary to retain ballistic properties. *(Chris Gibson)*

and maximum separation from the escaping carrier-plane.

As with Red Beard, twin-air turbines provided electrical power. Yellow Sun incorporated an electronic neutron initiator known as Blue Stone. Electrical power was provided for the fusing mechanisms by lead-acid accumulators in the tail of the Blue Danube casing and consisted of 6V motorcycle batteries procured from local suppliers.

Mk 1 held the Green Bamboo 'layer-cake' device and was the largest fission bomb deployed. It was an evolution of the Violet Club device tested at Maralinga. Violet Club incorporated a high-yield fission package called Green Grass which was expected to have a yield of at least 400KT. In total 37 bombs were built but none were tested. An unboosted fission device, it had a sphere of highly enriched uranium which, upon compression by the firing of 72 symmetrically positioned explosive charges, would compress the core beyond criticality and detonate.

Green Grass was considered a particularly dangerous weapon in that it used a disproportionately large quantity of fissile material in excess of 70kg (150lb) and, for safety, 20,000 steel ball bearings surrounded the core. To arm the bomb a bung was removed before flight, causing the ball bearings to spill out over a period of up to 30 minutes, but problems with them sticking together in freezing weather (not unknown on Lincolnshire airfields in winter) was challenging, while the rubberised lining tended to perish and decay. The time required to arm the bomb before take-off, and the limited warning time in the event of a surprise attack, rendered the entire system totally inadequate as a retaliatory weapon.

As a development of Violet Club, Yellow Sun 1 incorporated the Green Grass device but with much-improved safeing and arming, some 133,000 smaller ball bearings being used instead with a different retention system. Again, the casing was blunt-nosed to incur high drag and provide better separation from the carrier-plane than was possible with Blue Danube. But Yellow Sun Mk 1 was always considered an interim device pending the availability of a true thermonuclear fusion bomb. That possibility came with the cooperation afforded by the United States in 1958.

A modified version of the American W28 thermonuclear warhead was used instead of the Granite bombs tested at Christmas Island and plans to adopt the wholly British design for Yellow Sun Mk 2, Blue Steel and Blue Streak were abandoned. Renamed Red Snow, it was fitted to the Yellow Sun casing. While weighing much less than the Mk 1, with the addition of ballast it retained the mass of 3,290kg (7,250lb). Red Snow had a yield of 1.1MT and was exclusive to the Victor and the Vulcan, entering service in 1961 through to 1971, with a maximum deployed stockpile of 150.

The Red Snow warhead was manufactured in the UK from drawings and blueprints of the W28 warhead supplied by the United States, with some changes to critical components to suit British preferences, to provide effective mating with RAF handling and servicing processes and to tailor them to carriage in British aircraft. Red Snow was reduced in size and fitted with a smaller and much developed primary that was intended as a successor to Red Beard. Called Una, it was reduced in diameter and known as Ulysses and was intended for the Skybolt air-launched missile.

The design and manufacturing choice of UK atomic and thermonuclear weapons was in large part a result of economics. The dirty fission bombs prior to this point used highly enriched uranium, a decision based on price, which, at £19,200/kg was much cheaper than plutonium at £143,000/kg. Despite the higher quantity of HEU required for a given yield compared to a plutonium bomb, in 1960 each HEU weapon cost £1.56 million less that it would have using plutonium. With 37 weapons assembled, of which five were built as Violet Club, that decision saved the country almost £58 million (or £1.24 billion in 2017 money).

A shortage of plutonium was also the reason for the UK's decision to build dirty uranium bombs, a paucity made worse by a barter agreement with America to supply plutonium in exchange for uranium. But here too the UK was on the back foot, with only 860kg of HEU supplied by 1958 (compared to 472.2kg of plutonium by this date). With a less costly production process, the US was able to provide the UK with about 7,000kg of HEU that was purchased cheaply in return for the

sale of expensive plutonium, much preferred by the Americans.

All this changed with the revoking of the McMahon Act in 1958. Between 1960 and 1975, RAF Bomber Command was supplemented by up to 48 B28/43/57 warheads supplied by the Americans under a dual-key arrangement. Thus did the RAF maintain a stockpile of almost 200 bombs – the number originally determined to be the optimum for deterring Soviet aggression.

Blue Steel

The development of the Blue Steel stand-off weapon was long and tortuous, and its involved and serpentine emergence from a plethora of proposals and design concepts has already filled copious volumes and been described in numerous technical papers now freely available to the general public. Suffice it to say here that it grew out of recognition from the early 1950s that any aircraft penetrating Soviet Union air space would probably not survive long enough to carry out its mission

ABOVE Yellow Sun alongside a Victor B.1, showing the scale of the device and its bluff nose. *(Chris Gibson)*

BELOW Avro Vulcan B.1A XA903 with a Blue Steel trials round, probably taken at the company's Woodford facility, the aircraft in anti-flash white – a portent of its future stand-off role. *(Via Terry Panopalis)*

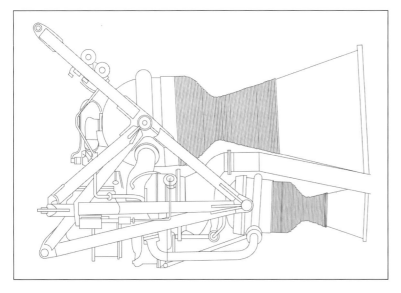

LEFT Propulsion for the Blue Steel was provided by the liquid propellant Armstrong Siddeley Stentor rocket motor, which had a main boost chamber delivering a thrust of 106.8kN (24,000lb) and a cruise motor with a thrust of 26.7kN (6,000lb). *(Chris Gibson)*

CENTRE The cutaway of the Stentor shows the two expansion chambers with the injector plates clearly visible at the back of the thrust chambers. The main motor would shut down after 29 seconds, followed by sustained operation of the cruise motor. *(Via David Baker)*

and certainly would never be seen again this side of the Iron Curtain.

It was necessary, therefore, to have the bomb released at some considerable distance from the target, allowing the parent aircraft to turn for home. In September 1954, OR 1132 defined such a stand-off weapon, which could propel itself by rocket motor at Mach 3 or greater to deliver a thermonuclear device to its target. Various proposals ensued and on 4 May 1956 the Ministry of Supply issued Avro with a contract for what would become known as Blue Steel, although Handley Page had placed a credible bid that was let down only by the inaccuracy of its proposed Elliott gyro-based guidance system.

Power for the missile was to be provided by a double-chamber Armstrong Siddeley Stentor Mk 101 rocket motor operating on propellants of hydrogen peroxide and kerosene (a coarse-grain paraffin). The engine operated on the principle of a main chamber providing a thrust of 106.8kN (24,000lb) for the boost phase followed by operation of the smaller, 26.7kN (6,000lb) thrust chamber for sustained cruise at Mach 2.3.

When deployed operationally, the phased propulsion called for the Blue Steel to separate

LEFT Utilising high-test peroxide (HTP) and kerosene propellants, much of the design drive to develop this type of motor came from captured German research activity and from a prolific series of motor designs using these propellants during the war. *(Via David Baker)*

from the V-bomber at 12,200m (40,000ft), free-fall to 9,750m (32,000ft), ignite the main chamber and climb to 18,000m (59,000ft) on the power of the main stage before cutting to the small stage for the final climb to maximum altitude and cruise to the target, whereupon the motor would cut out and the missile begin a terminal fall to its destination, usually for a pre-programmed air burst.

Blue Steel had a maximum range of 240km (150 miles) and a maximum altitude of 21,500m

RIGHT The trajectory of the Blue Steel after release at 9,750m (32,000ft) followed by a climb to 17,980m (59,000ft) at Mach 2.3, followed by shutdown and a steep dive to the target. *(Chris Gibson)*

BELOW Several proposals were made for advanced versions of Blue Steel that were to succeed the Skybolt era. Z.108 and Z.109 were longer-range missiles and Z.122 was to have been carried by the TSR.2 while Z.107 had extended range through a conformal tank in the ventral position. *(Chris Gibson)*

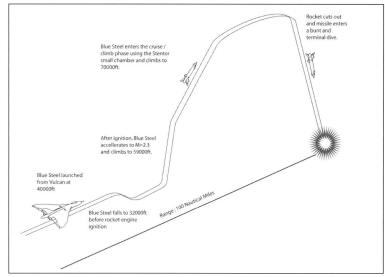

Rocket cuts out and missile enters a bunt and terminal dive.

Blue Steel enters the cruise / climb phase using the Stentor small chamber and climbs to 70000ft.

After ignition, Blue Steel accellerates to M=2.3 and climbs to 59000ft.

Blue Steel launched from Vulcan at 40000ft

Blue Steel falls to 32000ft before rocket engine ignition

Range : 100 Nautical Miles

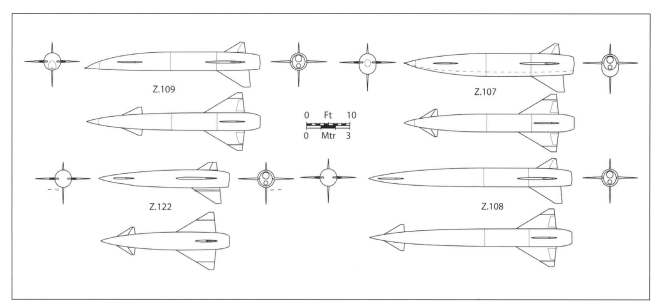

Z.109

Z.107

0 Ft 10
0 Mtr 3

Z.122

Z.108

(71,000ft), in speed and height well outside the capability of anti-aircraft missiles anticipated for the next several years. Of course, outright range could be traded for redefined flight paths optimised for a specific mission. Some profiles would have the main chamber reignite to accelerate the final part of the flight path to Mach 3, for instance when approaching very heavily defended targets.

The missile itself had a length of 10.7m (35.1ft), a maximum diameter of 1.22m (4ft) and a weight of 7,700kg (17,000lb) and as such

ABOVE This example of the aft end of a Blue Steel missile displays the two Stentor expansion chambers with integrated nozzle and aft plate.
(Via David Baker)

RIGHT Throughout the troubled Blue Steel development period, Avro put up several alternative options for a range of missile designs using combinations of different propulsion systems, an activity which some in the Air Staff felt was taking attention away from their primary job.
(Chris Gibson)

Blue Steel 1A / Z.81
Blue Steel 1* / Z.74
Blue Steel 1B / Z.84
Blue Steel 1D / Z.86
Blue Steel 1C / Z.85
Blue Steel 1*E / Z.87
Blue Steel 1*KG /Z.88

HTP   Hydrazine   90% HTP   Kerosene   Solid Boost

RIGHT An ageing Blue Steel case at Midland Air Museum, where the past lives on decades after the missile's retirement.
(Via David Baker)

LEFT Development of a supersonic strike capability in the form of the TSR.2 prompted new design concepts for weapons delivery, the futuristic concept based around deep-strike survivability from low-level intrusion and sophisticated countermeasures. Seen here is a surviving TSR.2 airframe at Duxford in 2006. *(Via David Baker)*

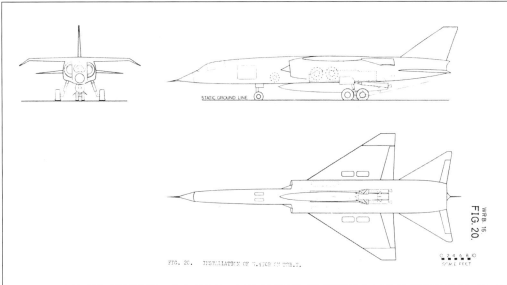

LEFT Designated W.170B, this Blue Steel evolution for TSR.2 was to have been powered by two large rocket motors which could have provided a range of 883km (549 miles) and also been carried by the Vulcan B.Mk 2. *(Chris Gibson)*

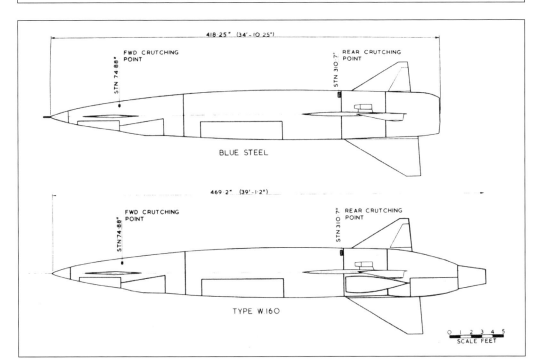

LEFT Evolving from a requirement for a low-level delivery weapon that could be carried by the TSR.2, Avro came up with the W.160 (bottom), shown here compared to the standard W.100/105 and with its Bristol Siddeley Viper 24 turbojet which would have had a thrust of 17.8kN (4,000lb), fed by air intakes under the wings. *(RAF Museum)*

was not very different in loaded mass to the compensated Yellow Sun as a weapon load to the carrier-plane. In service from 1962 until the end of 1970, no more than 50 were operational at any one time, with no more than 40 deployed on Victors and Vulcans.

The weapon was troublesome, due to the vagaries of its cooling system and the guidance equipment. Also it was highly complex to prepare for flight – an activity requiring at least seven hours – and the overall assessment within the RAF was that no more than half would perform satisfactorily in the event of war. Moreover, by the time it entered service the air defence threats of the Soviet Union were much greater, forcing the operational planning to embrace low level approach followed by pop-up to 300m (1,000ft) before releasing it in an adaptation of the toss-bomb technique.

While the airframe was a wholly British product from beginning to end, the weapon

LEFT This evocative shot of the unique design of the Handley Page Victor B.2 shows the nose-down carriage attitude of the Blue Steel weapon, the only aircraft other than the Vulcan ever to carry this missile. (Chris Gibson)

RIGHT Vulcan B. Mk 2 XL317 of 617 Squadron with a Blue Steel drill round, identified by its pale blue colour. Trial rounds were painted black while operational rounds were white. (Terry Panopalis)

RIGHT A Vulcan B.Mk 2 bombed up with an operational Blue Steel missile while ground crew brave the rain to prepare the aircraft for flight. The 7,257kg (16,000lb) missile had to be released between 93km (58 miles) and 204km (126 miles) from the target on an azimuth of no more than +/-40°. *(Kev Darling)*

it carried was a direct product of the 1958 agreement between the UK and the United States, revoking much of the McMahon Act. Blue Steel incorporated a Red Snow thermonuclear device derived from the W28 which was also used in the free-fall B28 bomb and in the AGM-28 Hound Dog (which see). The device itself had a length of 1.52m (5ft), a diameter of 50.8cm (1.67ft) and a weight of 760kg (1,675lb). The weapon was scaleable in yield although the RAF version had a fixed value.

WE.177

Since the demise of Red Beard in 1971, British bombs have used US thermonuclear devices adapted for use in UK casings and fittings. But the successor to Red Beard and Yellow Sun was the WE.177, which owed its origin to a meeting between premiers Macmillan and Eisenhower in May 1960 when an agreement was signed for the UK's purchase of 144

RIGHT Vulcan crew practise a scramble, typical of the high alert state not infrequently imposed on RAF Bomber Command stations operating the V-force. *(Via David Baker)*

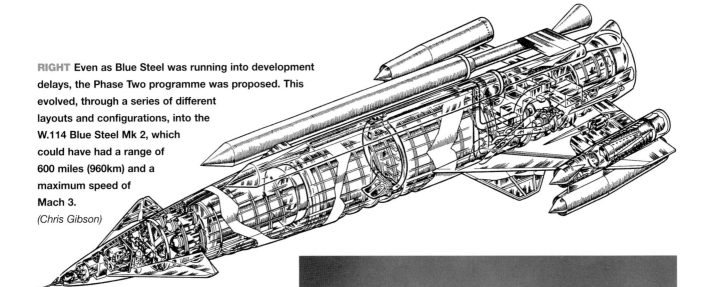

RIGHT Even as Blue Steel was running into development delays, the Phase Two programme was proposed. This evolved, through a series of different layouts and configurations, into the W.114 Blue Steel Mk 2, which could have had a range of 600 miles (960km) and a maximum speed of Mach 3. *(Chris Gibson)*

RIGHT Toward the end of the 1950s the US Air Force sought a stand-off missile to give its manned bomber force enhanced survivability. From this emerged the GAM-87A Skybolt, which in March 1960 was offered to the British for carriage on the Victor and Vulcan, the latter seen here visualised with two missiles under each wing. *(RAF)*

ABOVE Avro Vulcan B.Mk 2 XH537 undergoes flight trials with a definitive Skybolt GAM-87 missile under the starboard wing and the Delta 2 design under the port wing, promising to give the V-force greater range and flexibility above and beyond that provided by Blue Steel. *(RAF)*

LEFT By the early 1960s the British and American deterrents were converging through common hardware. Following on from the use of American designs for British nuclear weapons, Skybolt was seen by both air forces as the saviour of the manned bomber, previously threatened by increasingly formidable Soviet air defence systems. Here, a B-52 carries four dummy Skybolts. *(USAF)*

Douglas Skybolt stand-off missiles and the provision of American W59 warheads, which would be re-designated RE.179 with just a few modifications and changes in British service.

As work proceeded on Skybolt, the British opted to develop a successor to the tactical, free-fall Red Beard from the W59 warhead, producing a weapon system much smaller in size and approximately one-third as heavy. When Skybolt was cancelled in 1962 the Americans agreed to sell Polaris SLBM rockets to the UK, but its W58 warhead was rejected on technical grounds. The British opted for development of the RE.179 into a new warhead known as ET.317. Until Polaris became operational, the basic WE.177 was adapted into a slightly longer version, the WE.177B.

The British were concerned that the Polaris warhead had a relatively slow speed to the point of impact and there were concerns about anti-ballistic missile capabilities and the vulnerability of the warhead to X-rays from nearby explosions and in neutrons causing the primary to fizzle away and not detonate. This began the Chevaline programme (which see) which replaced one of the three warheads with decoys and for the redundant device to form the primary for a WE.177C fitted in casings for the WE.177B and ballasted accordingly so that it retained the same carriage dynamics and free-fall characteristics.

The WE.177A was a fission device with a variable yield of 0.5KT or 10KT, in a case that weighed 272kg (600lb) and had a length of 2.86m (9.4ft) and a maximum diameter of 43.3cm (1.42ft). The lower yield capacity was used in the Nuclear Depth Bomb (NDB) role with planned detonation above 40m (130ft) specifically in shallow coastal waters.

Approximately 107 were deployed between 1969 and 1992 for use variously between Victor, Valiant, Vulcan, Sea Vixen, Buccaneer, Jaguar, Sea Harrier FRS1, Wessex, Lynx, Wasp

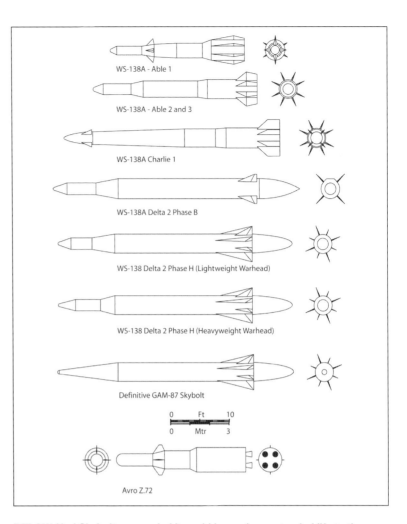

WS-138A - Able 1

WS-138A - Able 2 and 3

WS-138A Charlie 1

WS-138A Delta 2 Phase B

WS-138 Delta 2 Phase H (Lightweight Warhead)

WS-138 Delta 2 Phase H (Heavyweight Warhead)

Definitive GAM-87 Skybolt

0   Ft   10
0   Mtr   3

Avro Z.72

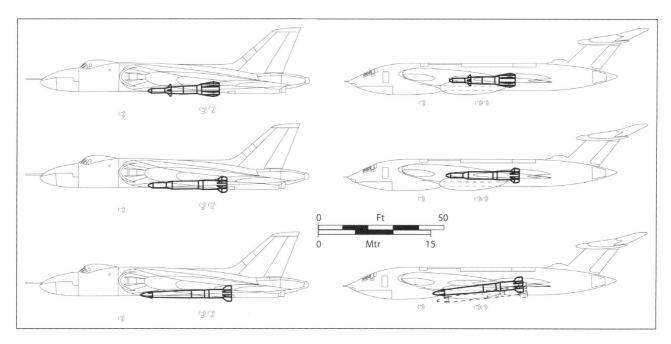

ABOVE Configurations of Skybolt initially proposed for both Victor and Vulcan, but adaptation to the latter proved easier than for the Victor, which was dropped as a candidate carrier. *(Chris Gibson)*

LEFT An operational WE-177A round sectionalised to show the internal arrangement of warhead and systems. *(Via David Baker)*

BELOW LEFT A WE.177 training round at the RAF Museum, Cosford, the definitive British nuclear weapon with wide applications for both air and naval deployment. *(Via David Baker)*

BELOW The safety and arming keys for the WE.177. *(Via David Baker)*

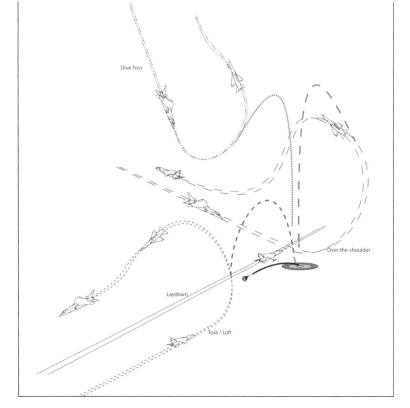

and Nimrod platforms. About 43 were deployed aboard vessels of frigate size and above for use by helicopters and Ikara systems as anti-submarine devices. It was also to have been carried by the cancelled Hawker P.1154.

With a length of 3.38m (11.1ft) and a diameter of 43.3cm (1.42ft), the WE.177B weighed 457kg (1,007.5lb), had a fixed yield of 450KT and was capable of airburst, impact or laydown deployment. With the transfer of the strategic deterrent to the Royal Navy, the RAF used the WE.177B in a tactical role with the Victor, Vulcan, Canberra and Tornado and potentially was to have been carried by the cancelled TSR.2. About 53 were produced but all had been retired by 1998.

With identical dimensions and weight as the WE.177B, the C variant had a fixed yield of 190KT and had been developed specifically to destroy airfields and other medium-size targets, a yield size which had been determined by NATO as optimum and the largest that should be employed for battlefield use to minimise collateral damage. About 60 were produced and most were deployed with RAF Germany in the tactical support role, the Navy retiring theirs in 1992 and all being retired from the RAF by 1998.

A wide range of possibilities existed for the further development of the type, which for nearly 30 years had been the primary tactical nuclear weapon in the British arsenal. One application studied but never implemented was to adapt the WE.177 into a submarine-launched torpedo weapon much like the Mk 24N Tigerfish nuclear armed torpedo, which was cancelled. Nevertheless, a 10KT warhead detonating at a

BELOW Buccaneer S1 XN929 bearing a white anti-flash finish for its nuclear role as a shipboard strike aircraft. *(Via David Baker)*

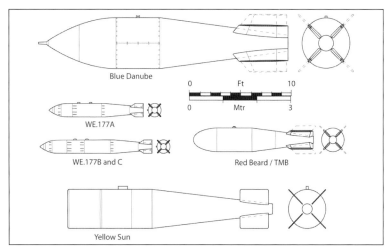

running depth of 12m (40ft) would have sunk a submarine at a depth of 600m (2,000ft).

Polaris

Development of the warhead for the American Polaris SLBM began immediately after the decision on 10 June 1963 to transfer the strategic nuclear deterrent from the RAF to the Navy. Polaris had been developed in the late 1950s as the third leg in the US strategic nuclear triad of land, sea and air systems

capable of striking the Soviet Union with powerful thermonuclear weapons. The first operational A-1 missile had a range of 1,900km (1,180 miles) and carried a single Mk 1 re-entry vehicle with a 600KT warhead. The A-2 had a range of 2,800km (1,740 miles) and the A-3 variant had a range of 3,700km (2,300 miles).

The British elected to buy the A3T version with a British ET.317 warhead, as described above under the WE.177 programme. It arose from the availability of the American W58 but with a distinctly different and new British-designed device based on the Cleo boosted-fission bomb, which had been tested in Nevada under the PAMPAS and TENDRAC test sequences. It was from this basic design that the warhead for the WE.177 was developed.

In design it used what is referred to as the fission-fusion-fission configuration where a boosted-fission device, *Jennie*, ignites a fusion secondary with the code name *Reggie* which derived from the fusion secondary of the W59. That, in turn, was encased in U-238 depleted uranium. Due to the enormous quantity of free neutrons liberated by the fusion event, unusually the U-238 fissions and allows a smaller but much cheaper bomb to be built than one which has a fission-fusion reaction. But this is a much dirtier bomb, despite the U-238 being a normal waste product of the manufacturing process for highly enriched uranium.

As befits a small missile inside a submarine, the internal dimensions of the Mk 2 re-entry vehicle, which the UK procured instead of the Mk 1, are small, with the diameter of the ER.317 being no greater than 23cm (9in), which is considerably smaller than the conventional W59 at 41.4cm (16.3in) for the W59 used in US weapons. Each missile carried three re-entry vehicles. A total of 144 Polaris missiles were delivered to the Royal Navy and 49 were fired in tests, most of them off the coast of Florida as part of the missile test range near Cape Canaveral. About 150 warheads were produced.

The first Royal Navy submarine equipped with Polaris, HMS *Resolution* set sail on its first patrol in June 1968 and was followed by HMS *Repulse* in June 1969, HMS *Renown* in August 1969 and HMS *Revenge* in September 1970. Faslane Naval Base, Scotland, was the home port for the nuclear-powered and Polaris-armed vessels and the missile system endured with the Royal Navy until replaced by the Trident submarines, with retirement of the British Polaris fleet between 1992 and 1994 matching the phased introduction of its replacement by *Vanguard*-class submarines.

Chevaline

By the early 1970s key decisions were necessary for the British nuclear deterrent. The Americans were replacing Polaris A3 with the longer-range Poseidon C3, each capable of sending between 10 and 14 warheads, each with a yield of 40KT, across a distance of 5,900km (3,665 miles). Developed during the 1960s, Poseidon was technically more advanced than Polaris and it began to replace the first generation of SLBMs from March 1971.

Long before then the British government was faced with the growing threat of Soviet anti-ballistic missile systems. With each Royal Navy submarine carrying 16 Polaris missiles, and each missile carrying three warheads, a single boat could release 48 nuclear devices. However, each triple-cluster on each Polaris would have little spread on re-entering the atmosphere and while descending upon the target the Russian ABM system could readily destroy all three with a single thermonuclear detonation high in the atmosphere as they streaked toward their target.

The Americans responded with the Poseidon missile, which could carry a greater number of warheads and stood a much greater chance of overwhelming Soviet defences and delivering the majority of its warheads to their respective targets. Poseidon was first test-fired on 16 August 1968 and moved rapidly through its test phase and initial deployment. It had one significant advantage over the triple re-entry vehicles of Polaris: multiple independently targeted re-entry vehicles (MIRVs), enabling each warhead to fly to a different target within range.

This feature was attractive to the Royal Navy. Instead of a single missile sending three warheads to a single target with a fan-spread, for instance to devastate Moscow, each Poseidon warhead could fly to a separate target. In this way, a single submarine equipped with Poseidon could address 224 targets instead of 16. The British government knew that

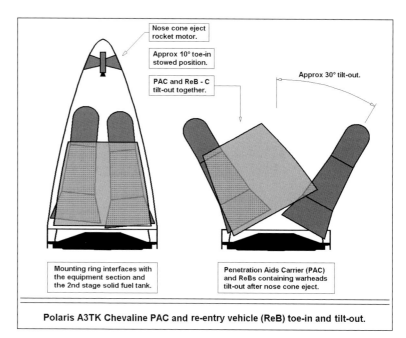

Nose cone eject rocket motor.

Approx 10° toe-in stowed position.

PAC and ReB - C tilt-out together.

Approx 30° tilt-out.

Mounting ring interfaces with the equipment section and the 2nd stage solid fuel tank.

Penetration Aids Carrier (PAC) and ReBs containing warheads tilt-out after nose cone eject.

Polaris A3TK Chevaline PAC and re-entry vehicle (ReB) toe-in and tilt-out.

the Russians had deployed approximately 100 ABMs around Moscow and that to decapitate the Soviet leadership it would require something radical. A one-year project was begun to find a solution, code name KH.793.

One option considered was to build two more submarines. With a four-boat fleet only a single submarine would be on station at a given time, with one returning from patrol, one in refit and one being readied for the next sailing. With five boats, two could be ready at all times somewhere within range of Moscow, ready to deliver 96 warheads – the 'Moscow Card' as it became known to the inner circles of government and the Ministry of Defence.

Another option was to use hardened warheads capable of resisting the electromagnetic pulse of a nuclear detonation plus penetration aids (penaids) to assist with confusing the enemy, together with decoys to confuse radar controllers and automated alert systems suddenly saturated with scores of seemingly viable nuclear warheads coming in. There was a still more sophisticated answer: to develop a British MIRV system for the Polaris missile, reducing to two the number of warheads carried so as to make room for the penaids and decoys.

But the Royal Navy had what they thought was a simpler and more certain solution: buy the American Poseidon with its MIRV cluster of 10–14 warheads and achieve commonality of

ABOVE The Chevaline A-3TK re-entry vehicle was developed for upgrading the existing British nuclear deterrent without switching to the American Poseidon missile. It had a penetration aids carrier (PAC) and two warheads, each of 225KT yield.
(Via David Baker)

RIGHT The entire
operational Chevaline
deployment sequence
is presented in
sequence according
to numbered stages,
designed to evade
Soviet anti-ballistic
missile systems.
(Brian Burnell)

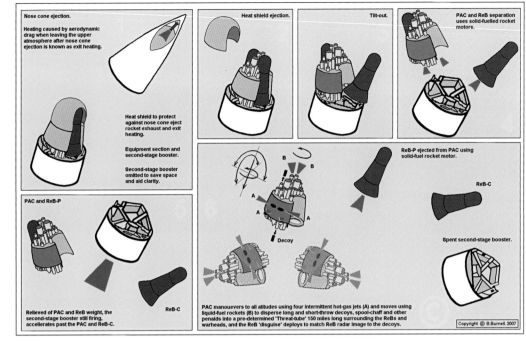

components, faster service introduction and, arguably, a cheaper and quicker response to the growing threat. Moreover, the Americans did not like the penaids/decoy solution, simply because being lighter than the incoming warheads they would slow appreciably and declutter the radar screens, thus facilitating the very conclusion they sought to avoid.

As the British veered away from buying into Poseidon, they focused on the Polaris A3TK and followed a possible mini-Poseidon solution, placing fewer warheads on the missile but with MIRV capability. Either way it would require development of a MIRV 'bus' to manoeuvre the separate warheads to individual targets.

The higher levels of decision-making decided to procure the Polaris A3TK, which would ideally replace the A3T, using a new bus with penaids and decoys and two manoeuvrable warheads, giving each boat the capacity to hit 96 targets. In administrative and bureaucratic logic, it was the same as buying an additional submarine but at a much lower cost – in theory. But the more advanced MIRV technology would be too expensive and the additional weight would require a larger and more powerful missile, so negating the advantage of that system in the first instance. So a compromise was necessary.

The decision was made in late 1973 under the government of Edward Heath, with full-scale go-ahead in January 1975, for a system preserving the single target approach but with greatly enhanced survivability. It had been planned to announce this to the public but an election got in the way and that never happened.

But another decision had been made – what to call the new system, a hybrid of the US concept. Initially the American system for penetration aids/decoys and MIRVed warheads went under the name 'Antelope', and the Ministry of Defence had thought to call the British adaptation 'Super Antelope', but an official was asked by the Secretary of Defence, Lord Carrington, to come up with a name that would disassociate the two projects and allay prying eyes and ears from linking the two. He called London Zoo and was told that there was a South African animal, like an antelope, known as a Chevaline – and so it became.

The British Chevaline project was one of the most secret in post-war defence development activities and would remain secret through several changes of government, its existence being denied by several Prime Ministers and only revealed in 1979 by Secretary of Defence Francis Pym during the Thatcher administration. With enormous cost overruns and profound technical difficulties, the project was too far along to cancel, although it would soon be replaced by the Trident system. By this time the Americans were moving on from Poseidon, leaving the British nuclear deterrent significantly behind the technical progress being made in the United States.

The primary threat to the British deterrent was Russia's ABM-1 Galosh system, and while the UK's Polaris A3TK solution reduced the number of warheads from three to two it significantly increased the likelihood of these getting through to the target. It came with an especially hardened primary for the thermonuclear fusion device and a significantly hardened re-entry vehicle.

While the 'standard' Polaris carried three 200KT warheads, the A3TK had a single ET.317 warhead attached to the terminal stage of the missile, as had all three warheads previously, and a second warhead on the penetration aid carrier (PAC) plus decoys and penaids. This was not a MIRV system but one that focused attention on getting through to the target, and for that assurance each bus carried 27 decoys plus the two warheads targeting for a scatter-spread on a single objective. In this way a single submarine would present 551 apparent warheads to the enemy and virtually guarantee that the real warheads would get through.

The ET.317 warheads used the new *Harriet* boosted-fission trigger with the *Reggie* fusion secondary from retired material which, reused in the Chevaline warheads, increased the yield to 225KT each, as per the information here for the WE.177. Chevaline's warhead concept had been defined by Peter Jones at Aldermaston in 1969 and in August 1972 President Nixon gave authority for three US underground nuclear tests in support of this programme.

The first test was carried out on 23 May 1974 and the following year the British Aircraft Corporation was given a contract to coordinate design. The first development flight launch from a flat pad at Cape Canaveral took place shortly before midnight on 12 September 1977, followed by the first development launched from a submerged HMS *Renown* during the evening of 14 November 1980. Both this and the following test launch four days later suffered partial malfunction due to problems with the control systems.

During a standard flight trajectory for a range of 4,630km (2,875 miles) the warhead would reach an apogee of 1,020km (630 miles) before heading down toward its target and profiles such as these were evaluated on

several additional test flights in June 1983, a year after HMS *Renown* went on its first operational patrol with the A-3TK/Chevaline assembly. Complete conversion to this system in all four British submarines was completed in September 1984 and from 1986 a new and improved propulsion system for the A-3TK was installed. The last Chevaline missile system was withdrawn in early 1998.

Trident D5

The Lockheed Trident C-4 system (Trident I) was developed with improved range and entered service in 1979, a year after a successor to the British Polaris A-3TK had been sought under two studies secretly commissioned by the Callaghan government. At a meeting between the Prime Minister and President Carter in January 1979 it was agreed that the UK could buy Trident for a new *Vanguard*-class submarine fleet and this was endorsed by Margaret Thatcher at the end of that year. On 15 July 1980 the government announced it would buy Trident I but this was changed to the Trident II D5 in March 1982.

The British Trident D5 would use the American Mk 4 or 4a MIRV bus but with British warheads based on the W76 with a yield of 100KT. The Royal Navy operates a maximum of eight warheads on each missile but would usually equip most with between one and four,

ABOVE The Trident II D5 was procured by the UK to replace Polaris, originally equipped with the ET.137 warhead but updated to carry the Chevaline system, the first Trident system declared operational in December 1994. *(US Navy)*

the government having taken the unusual step in 1999 of announcing that each boat will carry a maximum 48 warheads, implying an average of three on each missile, with a subsequent announcement lowering that to 40 per boat. Each of the four *Vanguard* boats carries the same 16 missiles as the *Resolution* class.

The British purchased 58 Trident II missiles, with an additional ten for firing trials without warheads, and in October 2007 the government announced that it would retain fewer than 160 warheads for the Trident II system. But the government has declared that each missile could carry up to 12 low-yield MIRV weapons

of about 10KT and some missiles are capable of being modified for the sub-strategic role, utilising them for tactical support, a role withdrawn from the RAF with the retirement of the WE.177 but utilising devices evolved from that programme. The UK Trident entered service in 1994, four years after the US Navy declared it operational and from December 1995 HMS *Victorious* deployed with single warheads for the sub-strategic role.

The British procurement agreement was finalised between premiers Reagan and Thatcher and included a clause contributing 5%, or $125 million, to the overall research and development cost of $2.5 billion for the entire programme. Each missile costs about $70 million. In 2002 the US decided on a life extension programme to make the D5 relevant as a strategic deterrent until at least 2040 and to achieve that by using available, off-the-shelf components to replace costly, ageing elements.

Prime Minister Tony Blair informed the House of Commons on 4 December 2006 that it would join the life extension programme and that it would support the construction of a new *Successor* class of submarine to sustain the British deterrent for the next several decades. Parliament approved the decision on 18 July 2016 and on 21 October that year the name of the submarine class was changed to *Dreadnought*, the first of which is expected to enter service by 2030. Each boat will carry 12 Trident II D5 missiles with up to eight warheads each in a W76-1/Mk 4A re-entry body, within a declared operational maximum of 180 British warheads in the stockpile.

Since 1969 the Royal Navy has been sustaining a continuous at-sea deterrent (CASD) under Operation Relentless and is supported in that role by government planning until at least the 2040s. The total cost of the British nuclear weapons programme is about 6% of the annual defence budget, which was £38 billion in 2017/18, showing annual UK expenditure of less than £2.3 billion on the nuclear deterrent.

US warheads to Britain

The involvement of American nuclear weapons in the British armed forces began largely with the 1958 Programs of Cooperation in which

bilateral agreements were made whereby the US would provide nuclear weapons and support services to NATO allies, who would operate the delivery systems. US forces would maintain control of the warheads, artillery shells and bombs until a high-level order would authorise them to be handed over to their hosts for use.

The pact known as the Agreement for Cooperation on the Uses of Atomic Energy for Mutual Defense Purposes involving UK forces was signed on 3 July 1958. From that date US weapons would flow to the UK, reaching a maximum deployed stockpile of 400 by the end of the 1970s. But the involvement of American atomic weapons actually predated that by several years.

The Royal Regiment of Artillery was the only British force to deploy US weapons with the British Army of the Rhine (BAOR). From 1954 the US had furnished the British with 113 Corporal surface-to-surface battlefield missiles armed with the W7 warhead, the sole country allowed to field this US weapon. After participating in tests at White Sands Missile Range, the British deployed Corporal operationally – with US-controlled warheads – from 1957, with the 27th and 47th Guided Weapons Regiments, until February 1967, the British being the last to retire it.

The UK had planned to develop its own short-range tactical missile, the Blue Water, but abandoned that in favour of buying the American Honest John and later the Lance. This was the first US nuclear-capable missile and carried the W31 warhead. The British deployed 120 between 1960 and 1979. The Lance with its W70 warhead was first deployed with the British Army in November 1976 and the 50th Missile Regiment, Royal Artillery, based at Mendon, had charge of 20 reloadable launchers for a stockpile of 85 warheads until retired in 1991 at the end of the Cold War.

Under the custodianship of the 570th US Army Artillery Group, the Americans supplied nuclear artillery shells from 1960 until 1987, including 36 W33 shells for British 8in howitzers, which from the mid-1980s were grouped into a single regiment, the 39th Heavy Regiment with four batteries comprising 12 front-line M110A2 self-propelled guns. AWE Aldermaston began development work on a

British nuclear warhead for the 8in howitzer but this was never completed. All shells of this class were withdrawn completely between 1987 and 1988 due to the NATO decision in December 1979 to reduce the number of theatre and battlefield weapons.

The British Army took on the M109 155mm howitzer in 1965 and three years later these were equipped with W48 warheads, each with a yield of only 0.072KT. These are described in the US section. Work on an atomic demolition mine (ADM) began in 1971 and 50 were acquired by the British Army with W45 and B54, delivering a yield of 0.01–1KT, for operation by the Royal Engineers. These 'backpack bombs' would have been used to support battlefield operations and for blowing bridges or main arteries as well as mining out areas as required.

The American Lance was acquired by the British Army in 1976 and 1991 with 85 in the nuclear stockpile, supported by about 85 with the W70 warhead, which had a variable yield of 1–100KT. All US nuclear munitions were withdrawn from the British Army following President Bush's announcement of 27 September 1991 that all these US tactical nuclear weapons would be withdrawn and all had been returned to the United States by July 1992.

BELOW The simplified internal layout of the W88 warhead for the US Trident II D5 missile. *(USN)*

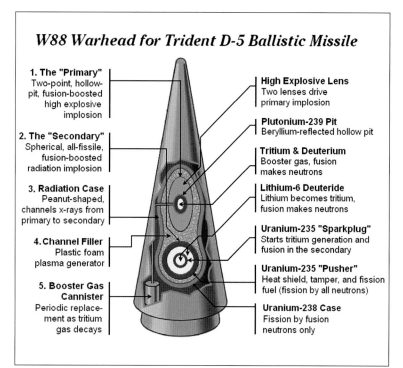

BELOW The simplified internal layout of the W88 warhead for the US Trident II D5 missile. *(USN)*

W88 Warhead for Trident D-5 Ballistic Missile

1. The "Primary"
Two-point, hollow-pit, fusion-boosted high explosive implosion

2. The "Secondary"
Spherical, all-fissile, fusion-boosted radiation implosion

3. Radiation Case
Peanut-shaped, channels x-rays from primary to secondary

4. Channel Filler
Plastic foam plasma generator

5. Booster Gas Cannister
Periodic replacement as tritium gas decays

High Explosive Lens
Two lenses drive primary implosion

Plutonium-239 Pit
Beryllium-reflected hollow pit

Tritium & Deuterium
Booster gas, fusion makes neutrons

Lithium-6 Deuteride
Lithium becomes tritium, fusion makes neutrons

Uranium-235 "Sparkplug"
Starts tritium generation and fusion in the secondary

Uranium-235 "Pusher"
Heat shield, tamper, and fission fuel (fission by all neutrons)

Uranium-238 Case
Fission by fusion neutrons only

Chapter Five

The French bomb

Determined to walk an independent path from Western alliances, France turned to the nuclear bomb as a way of protecting itself in the absence of a large standing army and built a triad of deterrent capabilities which placed it firmly in the forefront of nuclear power.

OPPOSITE The Dassault Rafale carries the air-sol moyenne portée-amélioré (ASMP-A) nuclear missile, the embodiment of France's grip on a strategic and tactical nuclear weapon system that its leaders have repeatedly said could be used against countries known to sponsor terrorism. *(Dassault)*

Immediately after the end of the Second World War France entered into a phase of rebuilding and reconstitution of national industries but lost no time in establishing the Commissariat à l'Energie Atomique (CEA), the Atomic Energy Commission. Set up on 18 October 1945, it came under the provisional government of General Charles de Gaulle as President. France itself is a proud nation with a long tradition in science, technology and engineering and had made the greatest stride in aviation before and during the First World War. Now, with peace settling once again over the nation, the country lost no time in restoring those capabilities and that tradition.

De Gaulle was particularly impressed with the atomic bombing of Hiroshima and Nagasaki

and recognised that this was a weapon of the future and that it was incumbent upon the new government to provide the country with its own independent deterrent and to restore France as a global leader in science and invention. France had been invaded three times in 70 years by German troops, who in that period had occupied the country for a total of ten years. The burning desire to never allow that to happen again, coupled to Gallic pride and sense of independence, framed the nuclear policies and plans France laid down from 1945.

Raoul Dautry and Frédérick Joliot-Curie had been appointed to manage the CEA and in 1946–47 they began recruiting personnel and establishing facilities for the development of atomic energy, with the aim of eventually producing nuclear weapons. A site for nuclear studies was chosen at Saclay, just south of Paris, and in July 1946 the organisation took control of an old fortress at the Fort de Châtillon, on the outskirts of the capital, where the first reactor was built. Known as the EL-1, it achieved criticality on 15 December 1948 and was used for research and training, producing very small quantities of plutonium along with artificial radioisotopes for experimental work in biology, medicine and industrial research.

Work was eventually transferred to the Saclay Nuclear Research Facility and a second one, EL-2, was opened as a heavy-water moderated uranium metal reactor. By mid-1949 the CEA had built a laboratory-size plutonium extraction facility at Le Bouchet, 40km (25 miles) south of Paris. By the end of the year it had produced the first milligram of plutonium in the form of purified salt and by the end of 1950 had extracted 10mg, with 100mg by the end of 1951. This was sufficient quantity for research purposes and a wide range of experimental data was obtained on the chemistry of plutonium and several key personnel were trained in health physics and remote-control techniques. This led to the significant improvement in chemical extraction methods and eventually to the use of tributyl phosphate as a solvent.

In parallel, a pilot processing plant had been set up at Fontenay-aux-Roses, where the first gram of plutonium was isolated from the spent uranium rods of EL-1 in 1954. By 1957 the reprocessing of rods in this way had yielded 200gm of plutonium, but the decision to build a bomb still required stimulus and that came from Francis Perrin, who in April 1951 became the new High Commissioner of the CEA, and Félix Gaillard, who was appointed Secretary of State for Atomic Energy. Gaillard would become Prime Minister in 1958 but in November 1951 he appointed Pierre Guillaumat administrator-general of the CEA.

These three men were instrumental in setting up a five-year plan that was approved by the National Assembly in July 1952. This called for the construction of plutonium production reactors and a plutonium extraction plant at Marcoule on the Rhône River. While it was not directly required within the plan, it was the implicit assumption that if a formal decision was made by the government to develop an atomic bomb all the infrastructure to achieve this would be in place. That prospect began to increasingly get the attention of the military, which created a Committee on Special Armaments, a euphemism for nuclear weapons.

The military were particularly interested in the use of atomic weapons on the battlefield and in understanding the effects of these weapons on their armed forces, so training centres were set up to provide courses in anti-radiation protection. Elsewhere, the political landscape was shifting. The Korean War of 1950–53 was a shock, and the defeat of French forces at Dien Bien Phu in French Indochina in 1954 struck a blow to France's self-esteem, which plunged to an all-time low. Probably for the wrong reasons, the French government decided that national pride demanded the development of a national atomic bomb, and that its international influence would be restored by achieving nuclear weapons capability.

But unlike the American, Russian and British bombs, France's nuclear capability would not emerge from the usual channels of executive and legislature, winding its way through institutional channels of approval and assent, but rather through consensus-building between interested parties occupying a wide range of influential and effective positions, under the continuous leadership and direction of the CEA. For several years the CEA worked through various channels of public communication, as

well as local and national leadership, to create a sense of inevitability and presented the possibility of atomic weapons as a fait accompli.

For three years the protagonists worked to create this sense of purpose and on 20 May 1955 a secret protocol was signed authorising the transfer of funds from the Armed Forces Ministry to the CEA, specifically for work on a French bomb. There had been no parliamentary debate and no formal discussion, but for the first time the French armed forces budget for

ABOVE Today the Saclay Nuclear Research Centre is part of the University of Paris-Saclay, the place where the Commissariat à l'Energie Atomique (CEA), the Atomic Energy Commission, was established. *(CEA)*

BELOW France's first nuclear weapon, the AN-11 was developed during the early 1960s. It demonstrated a yield of 60KT from a plutonium device declared operational from 1964. *(SNECMA)*

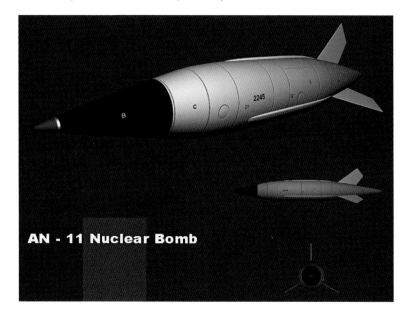

AN - 11 Nuclear Bomb

ABOVE The Sud Aviation Vautour (the one shown here is a IIB) was one of the first fighter-bomber types developed by France after the Second World War. Together with the Mirage IV, aircraft of this type carried France's first nuclear weapon. (Sud Aviation)

BELOW A Mirage IVP of 01.091 Squadron, a type equipped from early in its history with the capability of delivering nuclear weapons, deployed with the 300KT TN 80 from 1985–91. (Jerry Gunner)

1955–56 included funds for an atomic weapons programme. Other decisions, and international events, conspired to accelerate the programme and to expel any doubts about the need for such a device.

The Suez crisis of 1956 was pivotal in France's determination to turn its back on allies and proceed with an independent posture on defence and national security. A secret plan by France and Britain to wrest control of the Canal from Egypt's President Nasser had been rebuffed by Russia and by President Eisenhower who, in forcing a withdrawal of Anglo-French forces, humiliated both France and the UK. This provoked a close tie between the CEA and the military, who sought a way to provide the French government with the level of independence from allies and autonomy of action it sought to extricate itself from the nadir of international opinion.

A protocol was signed on 30 November 1956 tasking the CEA with preliminary design studies for an experimental atomic explosion, for preparing the scientific test itself, and for supplying all the necessary plutonium. This also called for the construction of a factory to separate and enrich the uranium. The armed

services would be responsible for preparations for the experiments and Colonel Charles Ailleret, the most outspoken advocate of a nuclear weapons programme, was made a General on 10 June 1958 and placed in charge of the Commandement des Armes Spéciales.

Already, on 11 April 1958, the last Prime Minister of the Fourth Republic, Félix Gaillard, had signed off the official order for the manufacture of an atomic bomb, which was to be tested during the first quarter of 1960. This decision was supported by the new government and marked with a substantial acceleration in preparation for an early test firing.

Once again, Charles de Gaulle was at the helm and now in the powerful position of President, and the very tenets of Gaullism were wedded to the concept of France's independent nuclear weapons programme. With the bomb, de Gaulle was further empowered to give France a more independent stance, ignoring the smaller nations and turning his back on America and, for that matter, to some extent the British, for whom he harboured a deep-seated contempt.

The first French atomic test took place on 13 February 1960 at a test site in Algeria, then a French colony. Named Gerboise Bleue, it was conducted from a device installed at the top of a 105m (344ft) tower south-west of Reggane in the Tanezrouft desert of Algeria. With an explosive yield of 60–70KT this plutonium device was three times that of the first tests carried out by the Americans or the British, and three more less powerful plutonium tests were completed by the end of 1961. The site had been selected in July 1957 and to prepare for it a number of French scientists and military personnel, including General Ailleret, visited the Nevada test site in the US.

The French tests created international protest, from neighbouring countries who feared the consequences of radioactive fallout and from the major powers who were, at this time, in a moratorium on atmospheric testing. Reacting, the French moved their tests underground, with 13 conducted by the end of 1966 in the Taourirt Tan Afella granite intrusion at In Ecker, a place also known as the Hoggar Massif. Detonations were conducted at the end of a spiral-shaped tunnel with safety doors at various intervals to reduce the amount of gas vented to the atmosphere.

The yield of the underground tests varied greatly, with a band of 3.6KT to 127KT. Budget books for the period 1960–65 clearly show that the tests were in support of a general plan to first provide an operational atomic bomb, the AN-11, with a yield of about 50KT which could be carried by the Mirage IV supersonic bomber. The first test of the AN-11 took place on 1 May 1962 and it was deployed operationally from 1964.

The tests had a broader objective too, for the French got caught up in the highly doubtful application of atomic explosions to civil engineering projects that required clearing of large amounts of earth for dams or hydro-electric schemes. The underground tests were a part of such general research, with many open-ended questions about the beneficial and peaceful use of this destructive force. But the tests came to an end in July 1962 when Algeria gained independence, the country no longer amenable to this activity continuing.

In late 1962 the uninhabited atolls of Mururoa and Fangataufa in the Pacific Ocean were chosen as locations where the French could resume atmospheric testing. Despite the Limited Test Ban Treaty signed on 5 August 1963, the French decided to go ahead with atmospheric tests of their own and between 1966 and 1974 some 46 detonations took place, all but five at Mururoa. Bombs were tested on towers, from balloons and dropped from the Mirage IVA. From 1965 to 1970 several warheads were tested for the thermonuclear devices designed for the IRBMs and SLBMs which by that time France had in development.

Preceded by two relatively low-yield precursor tests, the first two-stage thermonuclear bomb was tested on 24 August 1968. Code named Canopus, it had a yield of 2.6MT and was coincident with the start of the country's first uranium enrichment plant at Pierrelatte, opened in April 1967. The development of the French hydrogen bomb was largely due to the work of Roger Dautry at the Saclay laboratory of the CEA, who had been sought by the Ministry of Research and Atomic and Space Affairs. This initial detonation involved a device weighing 3,000kg (6,615lb) suspended from a balloon and detonated at 600m (1,970ft). It was followed on 8 September by a 1.2MT device.

Tests were planned around the evolving pattern of France's strategic geopolitical interests and in support of its military force planning and tactical requirements. The 1960s saw development of a nuclear submarine programme and also the introduction of tactical nuclear weapons. France went further than Britain in fielding ballistic missiles on land and at sea as well as maintaining a tactical bombing force. But it also went further than the diversified weapons assets of the UK.

During the first half of the 1970s, in addition to the operational deployment of silo-based and submarine-launched ballistic nuclear missiles, tests supported nuclear weapons being developed for the army Pluton missile, the Mirage IIIE tactical aircraft, the Jaguar A and the Super Étendard carrier-based strike aircraft. In addition, these tests were conducted under the cooperation of the United States, despite political differences. While initial efforts were completely independent of the United States, gradually throughout the later decades America aided France to a considerable degree in advancing its technology and improving missile design and warhead production.

ABOVE The Dassault Mirage 2000N was specifically developed to carry nuclear weapons, and 20 aircraft of this type currently support the airborne deterrent. This example appeared at the 2016 Royal International Air Tattoo at RAF Fairford, England. *(Adrian Pingstone)*

BELOW The air-sol moyenne portée-amélioré (ASMP-A) nuclear missile carried by the Rafale F3 fighter-bomber ensured a full tactical role for the French Air Force. *(Dassault)*

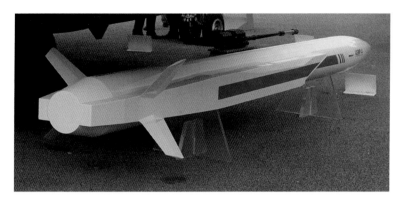

ABOVE The ASMP-A is deployed with French forces as a cruise weapon with a range of 500km (310 miles) and has a declared use to serve notice as a pre-strategic warning pending a full strategic retaliation, a unique negotiating posture not accepted by other NATO countries. The ASMP-A carries a TN 81 nuclear warhead with a yield of 100–300KT and is powered by a liquid ramjet engine. It can also be carried by the Super Étendard. (Aérospatiale)

RIGHT By the mid-1960s France had begun development of a silo-based ballistic missile system that was to be deployed in the Massif Centrale, the S2 being the first of its kind and with a 120KT warhead. The S2 had a range capable of striking deep into Russia and hitting Moscow from its launch silo in France. Eighteen silos were emplaced on the Plateau d'Albion before the S2's retirement in the mid-1980s. It was eventually replaced by the S3. (David Baker)

France had been a member of NATO since its formation in 1949 but the degree of cooperation with the US on ballistic missile design can be divided into three phases, the first running through the 1950s and 1960s when France was entirely independent, the second beginning in the early 1970s when an unprecedented level of cooperation and assistance was afforded by the US. During the mid-1980s the relationship entered its third phase, where France was able to reciprocate and match increasing American technology input with a reverse flow of equal veracity.

Aerospatiale sold missile technology to the American firm Hercules Inc, and France's

Société Européene de Propulsion (SEP) provided the US company Rocketdyne with detailed technical information sharing and signed a deal with that company and General Dynamics to provide exhaust nozzles for US rocket motors. The degree of cooperation between France and America is far higher in rocket propulsion and space-related technologies than between many other democratic countries. Reciprocating, French elements for its ballistic missiles were tested in the US and France supplied components for American missiles.

In some respects the relationship was a political ploy by American politicians to keep France close to them, in the knowledge that this country had equalled their own technical achievements and was not averse to sharing that capability with its own allies, including a technical deal to help Israel acquire a nuclear weapons capability (which see). This concerned the Americans, who felt they had to have France close by, and most of that manipulating and cosseting occurred during the tenure of President Nixon and Henry Kissinger.

All this largely came about after the Strategic Arms Limitations Talks (SALT) agreement of May 1972 was struck between the US and Russia, excluding France and the UK. It helped the West compensate for these limitations on the rate of increase in the arsenals of the US to stiffen the deterrent capabilities of the French. In looking in depth at France's nuclear arsenal, the US Defense Department assessed the credibility of those forces and concluded that there were certain inadequacies which could be addressed through stronger ties with America and an enhanced transfer of technological information on ballistic missile design, and this was the origin of the dramatic shift in that relationship.

This led to a long shopping list of items that the US was willing to provide for the French, including nuclear safety devices, new and improved long-range missile motors, unrestricted access to US computer technology and software systems, technology for MIRV warheads, and even assistance with France's nuclear-powered submarines. It took a while for Congress to agree, sessions being held in secret for discussion of these special agreements and deals, because the open hostility of de Gaulle to American forces on French soil had led to the expulsion of 70,000 US troops and a refusal to sign the Partial Test Ban Treaty of 1963.

Prior to this any cooperation between France and Britain was out of the question because of the degree of American assistance being received by Britain from the US, extending even to the provision of detailed drawings for the rocket motors for the Blue Streak LRBM. The Americans would not countenance any transfer by Britain of technology associated with the nuclear programme or missile projects for fear that France would get its hands on American know-how by the back door. It was this entrenched view of the status quo that proved a difficult obstruction when Congressional committees were asked to endorse a sudden switch in stance between America and France.

By the 1980s France had a developed, highly evolved and sophisticated arsenal of nuclear weapons, had already gone through several generations of modified and updated weapon systems (see next section) and had started deployment of a stand-off missile, achieving what the UK had sought to achieve with Blue Steel, with varying degrees of success. The end of the Cold War in 1991 brought radical change and the last silo-based missiles were disposed of as well as the tactical missiles Pluton and Hadès, the last of which was withdrawn in 1997. The sea-based deterrent remains in the form of four SLBM-equipped submarines and carrier-borne ASMP-A (air-sol moyenne portée-amélioré) nuclear missiles. The Air Force retains some ASMP-equipped Rafale N, deployed from 2010.

French tests

In total, between 13 February 1960 and 27 January 1996 France conducted 210 tests, with 193 in the South Pacific of which 80 were at the Mururoa site. Of these, 56 were underground and the rest in the atmosphere. About 90% of the atmospheric tests were carried out between 1968 and 1974 but between 1975 and 1996 some 48% of those remaining tests displayed yields only of 4KT to 15KT. In general, France carried out an average of 20 tests for development of each specific warhead or bomb,

compared to about six for the US. The total yield from all French tests accounted for 2.5% of all nuclear testing worldwide.

The French stockpile

France gradually expanded and developed its stockpile of nuclear weapons from the first deployment of four gravity bombs in 1964 with the Mirage IV to more than 100 by the beginning of the 1970s, more than 200 in 1976, exceeding 300 in 1985, 400 in 1987 and 500 in 1990. Since then, with an end to the Cold War, the arsenal has slowly declined until by 2008 it stood at around 300, since when it has stabilised at that level.

The first silo-based IRBM for France's land-based strategic deterrent was the S2 in 1971, a system known as the SSBS, or sol-sol balistique stratégique (surface-to-surface strategic ballistic missile). This system was retired at the end of the Cold War from 1991. Introduced in 1971 too, the MSBS or mer-sol balistique stratégique (sea-to-surface ballistic missile) has progressively evolved through increasing capabilities and is operational today

LEFT At 3,500km (2,200 miles), the S3 had a much greater range than its predecessor and continued to serve until 1991, each warhead carrying a nuclear package with a yield of 1.2MT.
(David Baker)

LEFT The two stages of the S3 missile can be seen clearly on display, the first stage being identical to the S2 but with a much-improved and more powerful second stage. *(David Baker)*

with four *Triomphant*-class submarines, each carrying 16 missiles.

In 2017 France has at least 280 deployed strategic nuclear warheads and at least a further ten in reserve, for a notional stockpile of 300.

AN air-launched series

The AN-11 was France's first nuclear weapon, developed from a precursor that was detonated on 13 February 1960, followed by a full test of the definitive AN-11 on 1 May 1962. A fission implosion type, the plutonium weapon had a yield of 60KT and a weight of 1,500kg (3,306lb) and was designed for high-altitude air-drop from the Mirage IV or, rarely, the Sud Aviation Vautour. It was deployed operationally between 1964–67, whence it was progressively replaced by the more compact 70KT AN-22 which had a weight of 700kg (1,540lb). The AN-22 was carried by 36 Mirage IV and retired on 1 July 1988 to be replaced by the ASMP.

The AN-52 was a low-yield tactical air-drop weapon which was based on the AN-51 warhead for the Pluton missile. First tested on 28 August 1972, it had a selectable yield of 6–8KT or 25KT, weighed 455kg (1,003lb) and was carried by the Mirage IIIE, Jaguar A and Étendard aircraft. Some 30 Mirage 2000N-K1 aircraft were also equipped to carry the AN-52, which was first introduced in 1972, retired in 1992 and replaced by the ASMP cruise missile.

Deployed as a pre-strategic cruise missile to provide a 'notice of intent' of the nation's willingness, under further attack, to escalate to a full strategic nuclear strike, the ASMP reached operational deployment on 1 May 1986 and is currently carried by 20 Mirage 2000N and Rafale F3 ground attack aircraft. Initially carrying the TN 80 warhead, the ASMP has a range of 300km (500 miles) and had been under development since 1974. With a yield of 300KT, the TN 80 weighed about 200kg (441lb) and has been in service since 1988, but is being superseded by the ASMP-A, which entered service in 2009.

ABOVE The solid propellant first stage has four exhaust nozzles, the design and choice of propellant supporting a rapid response time with little delay between warning and launch. *(David Baker)*

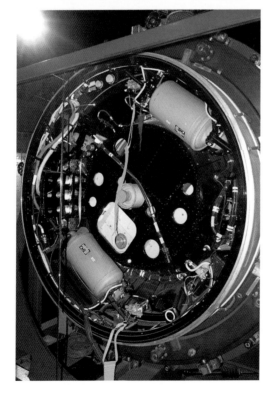

RIGHT The warhead compartment on the S3 attached to the top of the upper stage where the TN 61 thermonuclear warhead was carried. *(David Baker)*

RIGHT The inertial measurement unit for the S3 was common to several platforms, and the overall technology driven by the nuclear weapons programme helped France take the European lead in space launch vehicles. *(David Baker)*

Land-based warheads

The development of a land-based strategic missile evolved gradually between 1955 and 1965 during which official approval was granted in 1963. The first tests were performed in 1965 at the CEL facility near Biscarosse, when the initial design for the SSBS S-112 was launched on seven test flights towards the Azores and tracked by a station on Flores Island. Underground silos were used in these tests, which ended in 1967 and were succeeded by a second series using the S-01 between 1967 and 1968.

The SSBS S1 consisted of two P-10 solid-propellant rocket stages of similar thrust but was never developed, the S-2 being approved instead, consisting of the second stage of the S1 but with a more powerful first stage. The MR 31 was the warhead developed for the S2, France's first IRBM, which had a range of 3,000km (1,900 miles). The device had an explosive yield of 120KT and the missile entered service in 1971, with a maximum 18 silos in the Plateau d'Albion, before it was retired in 1984 in favour of the S3.

The first stage of the S2 and the S3 were powered by a P16 solid propellant rocket motor exhausting through four nozzles but the second stage on the S3 was improved, extending the range to 3,500km (2,200 miles). The S3 was equipped with the TN 61 warhead, which had a yield of 1.2MT and saw service between 1977 and 1991.

The SSBS missiles resided in silos 23.8m (78ft) deep and hardened to withstand an overpressure of 21kg/cm² (300lb/in²) so that they could ride out a nuclear attack at relatively short range. The 1.4m (4.6ft) thick concrete doors would move horizontally to expose the hot-launch fire tube. The missiles rested on a pedestal suspended from cable secured to pulleys and four cylinders on the floor. The fire

control centre was 400m (1,312ft) underground with two officers on duty at any one time.

Conceived in 1965, the Pluton tactical nuclear missile is a land-mobile system with a range of 10–120km (6–75 miles) and a CEP of 150m (500ft), developed as a replacement for the American Honest John missile. France had been host to Honest John since 1961 but decided to replace it completely with their own sol-sol balistique tactique (SSBT).

Test trials started with the first launch on 25 January 1968 and it entered operational service on 1 May 1974. The Army originally wanted 120 Pluton missiles but only 44 launchers were built, 35 of which were considered front-line nuclear-capable forces. Pluton carried the AN-51, which came in two versions of 10KT and 25KT yield. With an altitude-programmed fuse, the first test of the pure plutonium fission device occurred on 5 June 1971, with 70 warheads eventually being produced for the 35 Pluton missiles. The system was decommissioned by 1993.

In 1975 the Direction des Engins started work

ABOVE Until the mid-1970s, France used American short-range missiles such as Honest John, but development of the Pluton road-mobile tactical weapon gave a new capability, with 35 launch systems being built. *(David Baker)*

LEFT The *Redoutable* was France's first nuclear-powered submarine equipped to carry 16 M1, M2, M20 and M4 MSBS nuclear missiles. In commission from 1971 to 2008, the six submarines of this class have now been retired. *(Via David Baker)*

CENTRE The 'forest' of 16 launch tubes in pairs down opposing sides of the submarine hull from where a retaliatory strike would have begun. *(David Baker)*

on a second-generation land-mobile missile named Hadès which was to have been lighter, smaller and have a more effective warhead, in that it took advantage of Enhanced Radiation Weapon (ERW) technology to offer field commanders a 5–15KT weapon which could deliver intense bursts of lethal radiation while minimising the destructive effect, a neutron bomb in all but name. In December 1976 President Giscard d'Estaing declared that France had already tested a neutron bomb, but without specifying when and of what capability. It is believed that associated tests were of neutron-bomb relevant components and not a fully developed weapon. After the end of the Cold War the Hadès programme was placed on hold.

Submarine-based warheads

Development of France's sea-based strategic deterrent was, like the SSBS missiles, managed by Aerospatiale, with the first tests performed in 1966 from the Hammaguir launch complex in Algeria. These were followed with live firings from underwater rigs off Toulon. The first submarine launches were from the *Gymnote* in 1967 and continued at the CEL facility near Biscarosse in 1968. The prototype missile was fired from France's first nuclear-powered submarine, the *Redoutable*, in 1969

LEFT The author's wife Ann at the control station of the *Redoutable* from where the boat would be navigated and positioned at secret locations, undetected and on constant standby for the call to fire at predetermined targets. *(David Baker)*

RIGHT A scale drawing showing the cross-section of the new *Triomphant* class (right) with the M51 missile, compared with the M4 missile of the *Redoutable* class that has now been retired. *(SNLE)*

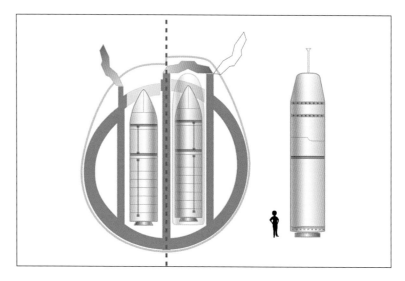

and the first operational M1 was installed in this boat in 1971, followed by *Le Terrible* in 1972. In this early phase, France commissioned six submarines of the *Redoutable* class, each equipped with 16 launch tubes.

The M1 had a range of 2,500km (1,550 miles), but an updated missile – the M2 – extended this to 3,000km (1,860 miles) and was installed in *Le Foudroyant* in 1974. A further variant, the M20, was installed in *L'Indomitable* in 1977. *Le Tonnant* and *L'Inflexible* were retrofitted with the M20 before the end of the decade. The M20 was equipped with either the TN60 or the TN61 warhead with a yield of 1.MT and penetration aids but was withdrawn by 1991. Excluding *Redoutable*, the fleet received the considerably more capable M4 missile with six TN71, 150KT warheads from 1985.

Each M4 had a range of 5,000km (3,100 miles) and provided the MSBS with its first MIRV capability, but the six-boat fleet was replaced by the four submarines of the *Triomphant* class, commissioned between 1997 and 2010. The last *Redoubtable* class boat had been retired in 2008. Capable of carrying the M45 missile with six 110KT TN 75 warheads and with a range of 6,000km (3,725 miles), the hardened warheads were of such advanced design that they prompted the French to resume nuclear testing for 1995 and 1996.

Considerable further development of the MSBS programme involved a planned M5 missile that was originally conceived with a range of 11,000km (6,800 miles) but which was adapted into the M51, with a range of more than 8,000km (4,975 miles), with a first flight test on 9 November 2006 followed by two additional tests in successive years. Operationally deployed from 27 September 2010, the M51.1 has between six and ten TN 75 warheads, each with a yield of 107KT, while the M51.2 has the new 150KT warhead. A new upgrade, designated M51.3, is expected to enter service in 2025. France is now working on a successor submarine class for service from 2035.

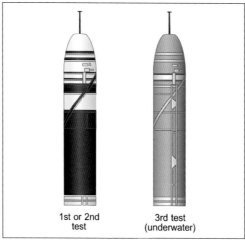

1st or 2nd test

3rd test (underwater)

LEFT The M51 missile in test configurations. Like the US Trident II D5, the M51 extends a spike after launch to part the air and present a streamlined flow around the otherwise blunt nose, compacted to a bluff shape to pack additional propellant within the allowable length. *(SNLE)*

BELOW An M51 is manoeuvred around prior to emplacement within a firing tube on a *Triomphant*-class submarine. *(David Baker)*

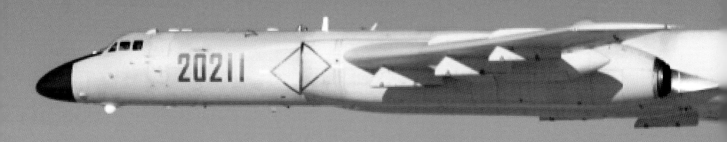

Chapter Six

The Chinese bomb

Urged on by a communist government intent on responding to the West's military capabilities, China sought help from Russia in building a nuclear nation, turning to atomic weapons and a more independent stance when it fell out with its neighbour.

OPPOSITE China has only recently developed a long-range nuclear strike capability, based around the H-6K which is a developed version of the Tupolev Tu-16 bomber. *(CNS)*

The political contest for the control of China that began in the 1930s was put on hold during the Japanese invasion and occupation until after the Second World War, but upon hearing about the atomic bomb and its effect on Hiroshima and Nagasaki, Deng Xiaoping made a firm commitment to develop such a capability for his country. After the seizure of power by the Communist Party, which took control of the nation in 1949, the essential building blocks for that development were put in place. Political and social imperatives prevented an immediate programme of nuclearisation because there was much infrastructure to build and consolidate after such a long period of conflict and war.

The spur to make a focused commitment came during the crisis of 1954–55 when a short conflict broke out between the People's Republic of China (PRC) on the mainland and the Republic of China which had fled to Formosa, now Taiwan. At first the Americans sided with the PRC, but after the outbreak of the Korean War on 25 June 1950 President Truman declared that it was in the interests of security that the strait between the mainland and the island should be a neutral zone, and placed the US 7th Fleet there as a hedge against inflated Chinese aggression.

This was the first real test for the West in dealing with a communist state and concerns ran high as China supported the North Korean invasion of South Korea. It was the perceived military threat that pressed Deng Xiaoping to formally start a nuclear weapons development programme. With US troops now stationed on Taiwan, the Chinese premier made direct reference to the effect Russia's atom bomb had on the West, believing that if his country had 'the bomb' it would elevate its status on the world stage. Aware that he would not be in a position to match the United States, or perhaps the Soviet Union, the advantage was implicit in the reaction to Russia's bomb.

In the early 1950s China had excellent relations with Russia and received considerable help in its nuclear aspiration. For the first decade the core of China's nuclear industry was put together by the Russians, while China itself harvested knowledge from its scientists and engineers who returned home after being educated in the West. Many had been schooled at American universities and worked in US laboratories. The communist incursion into South Korea, however, discouraged American enthusiasm for its Chinese workers and several were deported or simply sacked, returning to their native country with many brilliant ideas and concepts that had helped stimulate rocketry in the United States and would now do it again in China.

The return of China's expatriate workforce was encouraged with tempting offers of working on the nascent nuclear programme, much of which was a repeat of the initial steps taken by Russia a few years earlier. But China tended to group civil and military research into a single technological application, and much of the encouraging recruitment centred on discussion of building a nuclear power programme for China's accelerating demand for electricity and industrial infrastructure. The fact that it had dual branches only added to the interest shown by key players in the development of China's first bomb.

The Russians helped considerably, and without the training of China's new educated elite at facilities in the Soviet Union it would have been several decades before a Chinese bomb appeared. Several thousand personnel were sent directly from China to Russia, where Soviet technicians instructed them not only in the development of weapons-grade production plants but in the design and assembly of bombs and the technical merging with operational deployment.

Soviet help was formalised on 14 February 1950 when the two countries signed a 30-year Treaty of Friendship, and while the USSR supplied China with large quantities of material for their infrastructure China invested in national facilities and core research programmes, which would serve them well until the total breakdown of the relationship in 1960. From 1955 to 1958, however, the Russians virtually ran China's aircraft, weapons and nuclear programmes, but the origins of that relationship were already in place. On 28 September 1954 the Chinese Communist Party Politburo set up the Central Military Commission with sole responsibility for nuclear weapons policy and for command and control. That month Marshal Peng Duhai was appointed as the first minister of the Ministry of National Defence.

During the year relevant ministries reported to central government on the uranium ore situation and the decision was made to go full speed on development of the nuclear power industry, firmly aware that this would accelerate development of the nuclear weapons programme as well. In early January 1955 the noted physicist Qian Sanqiang briefed Premier Zhou Enlai and associates on the essential basic principles of nuclear weapons and on the ability of China to produce them. Mao chaired a meeting on the 15th with the Central Secretariat of the Politburo and approved a development programme, code named 02.

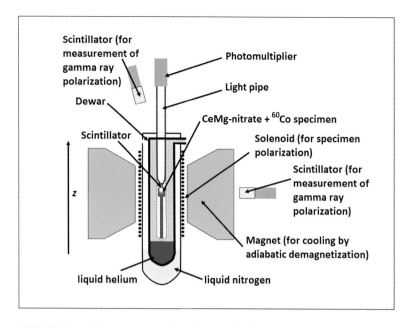

ABOVE One of Wu's most outstanding contributions was to study the physics of Beta decay, which had confounded experimenters since it was first published by Enrico Fermi in 1934. The experiment involved a copper sulphate to which was added a detergent to produce an even film to demonstrate that emitted electrons suspected of losing energy were in fact not doing that at all. The problem with experiments conducted previously was that there were errors in the manner in which they had been set up. *(Via David Baker)*

Two days later Russia publicly announced that it was to help China build a nuclear power industry, and a bilateral agreement to that effect was put in place on 27 April, the first of six Sino-Soviet

LEFT Hu Ning (1916–97) was born in China and educated in America but returned to his home country in 1951 to become a prominent figure at the Institute of Modern Physics, and was a researcher at the Joint Institute for Nuclear Research in the Soviet Union. He made major contributions to the Chinese nuclear programme and is remembered by this bronze bust in Beijing University. *(CNS)*

nuclear cooperation and assistance agreements made between this date and 1958. Meanwhile, on 20 January, China and Russia signed a secret protocol to the agreement for geological surveys carried out by Russian and Chinese scientists searching for uranium deposits, also allowing for the sale of such ores to the Russians. And at the end of the month the State Council accepted Soviet assistance in ratification.

A year later, with work well under way, 12 key tasks were included in the 12-year Long-Term Plan for the Development of Science and Technology. These embraced not only nuclear weapons development but also nuclear energy, rocketry and jet propulsion. The broader scope was undoubtedly influenced by the return of several rocket scientists who had been working in the United States but who were levered out for fear they would convey secrets to China while they worked in America. At least one well-known Chinese scientist had made a major contribution to early work at the Jet Propulsion Laboratory, Pasadena, California, before he was sent back.

Mao Zedong gave the nuclear programme his blessing on 25 April 1956 and reaffirmed the need for Chinese weapons, and on 25 May construction started on a Soviet heavy water research reactor and cyclotron at Tuoli outside Beijing. A day later the Central Military Commission established the Fifth Academy to start work on ballistic missiles, and a number of weeks later work started on a major rail line serving a missile test range that was to be set up at Shuangchengzi.

As part of a further agreement with Russia for assistance in building the nuclear research facilities, the USSR agreed to sell the Chinese two 8A11 missiles. Known in the West as the R-1, this was a copy of the German A-4 (V-2) rocket, which had been re-engineered by the Russians and introduced into service in November 1950. It could carry a 785kg (1,731lb) high-explosive warhead and had a maximum range of 270km (170 miles).

The Fifth Academy was founded in October 1956 and preliminary work started on a rocket manufacturing base and launch site and two months later the site surveying for uranium deposits was transferred to Chinese control with Russian help. Further links followed when Russia granted China a licence to manufacture the Tupolev Tu-16 medium bomber. This marked the start of a new and even closer relationship between the two communist countries and reached a peak when an agreement was signed on 15 October 1957 in which the Soviets agreed to supply China with a prototype fission bomb together with two 8Zh38 (R-2) rockets and all the associated technical data.

The R-2 was an evolution of the R-1 and had more than double the range through an engineering redesign that increased the length of the propellant tanks and incorporated significant refinements which greatly improved on the original V-2. This was the final Russian derivative of the German wartime rocket and it

Nuclear Test Site, Lop Nur, China, 20 October 1964

Ground Zero

entered military service in 1953, but it was not able to carry atomic weapons. Despite this the Russians placed great store on it as a battlefield weapon and produced a total 1,545 of both the R-1 and the R-2.

The R-2 was the design template for China's first experimental ballistic missile, code named the 1059 but popularly known as the Dongfeng 1, which had a range of 590km (366 miles). China also received some 8A61 ('Scud') missiles from Russia. These provided a solid and encouraging foundation for building an indigenous weapons industry based on nuclear weapons and ballistic missiles, and formed the basis of China's eventual emergence as a space-faring power in addition to its military technology.

On 11 February 1958, encouraged by the preparatory work, Deng Xiaoping approved several sites for an indigenous nuclear industry including a uranium enrichment plant, a weapons development site at Qinghai and three uranium mines at Chenxian, Hengshan Dapu and Shangrao. Confidently, he told an expanded meeting of the Central Military Commission on 21 June that he fully expected 'China to develop atomic bombs, hydrogen bombs, and intercontinental missiles within ten years'. One of the great strengths of the Chinese programme was the level of information absorbed by the leadership, which availed itself of the necessary technical information to make qualitative judgements.

Based on the pace of progress to date, in August 1958 the Central Committee authorised development of nuclear-powered submarine technology and that October a structure was in place for such a boat, specifically its reactor. But just as the transfer of technology and engineering information from the Soviet Union – now a rising space power – was reaching a peak, political fractures severed the links, and on 20 June the Soviet Union sent a letter notifying the Chinese Communist Party Central Committee that it would not, after all, be providing an atom bomb.

This news came a month after the first two Soviet Tu-16 bombers arrived at the Harbin Aircraft Factory together with one unassembled aircraft. That aircraft was quickly put together, redesignated Hong-6 (H-6) and put in the air for the first time on 27 September. In 1965 it

LEFT China has encouraged a feeling of national pride in its scientific and engineering accomplishments and proudly exhibits replicas of its early atomic weapons, although this particular exhibit looks very close to the American Fat Man design. *(Via David Baker)*

LEFT China began simultaneous development of ballistic missiles and nuclear weapons and adopted the strategy that a deterrent force to counter US military superiority was better achieved by long-range rockets. A start was made with the licensed manufacture of Russia's R-2 (SS-2 Sibling), essentially a copy of the V-2, albeit much modified and designated Dongfen 1 (DF-1). *(David Baker)*

These unique institutions were essential, as the Russian work force now began leaving the country in large numbers, the depth of Russo-Chinese cooperation having been significant. All had departed by 23 August and from this date there was no further contact for several decades. As a fillip to morale, China launched its first R-2 in September, followed on 5 November by the first 1059 missile based on the R-2. Nine days later Qian Xuesen started development of the Dongfen-3 (DF-3), a missile with a projected range of 2,500km (1,550 miles), placing himself as the chief designer.

Development of the missile had been influenced by Russia's 8K65 (R-14) more commonly known as the SS-5, an intermediate-range ballistic missile which began tests in 1959 and would eventually equip China's Strategic Rocket Forces, and which served as a template for the DF-3, the first stage of which was a direct copy. As developed, the Chinese missile was designed from the outset to carry an atomic warhead, later a thermonuclear device, with substantial design input from Tu-Shou'e and Sun Jiadong, with the missile being produced at Factory 211.

Meanwhile, the Chenxian uranium mine began partial operation in early 1962, followed by trial extraction from the Linxian Uranium Mine in Hunan Province in April. In Beijing, the first major report to the senior leadership from the Institute of Nuclear Weapons projected an initial test with a fission device in 1964. This was followed in August 1963 by the beginning of a suitable warhead design to allow the fission bomb to be attached to that missile.

would be used for the first air-drop of a Chinese fission bomb, but in the meantime, by the end of 1958, two production lines had been set up at Shenyang and Nancheng for the production of tactical missiles and rocket motors.

On 12 February 1960, with the pace picking up, the Uranium Enrichment Laboratory of the Institute of Atomic Energy, under the Chinese Academy of Sciences, was completed and turned over to production. Just seven days later China launched its first indigenous sounding rocket for ballistic flights, carrying instruments to the edge of space, while in Beijing the Institute of Nuclear Weapons started researching a purely home-grown fission bomb.

By the end of the year the Uranium Plant at the Jiuquan Complex had produced its first satisfactory uranium hexafluoride. This was followed on 14 January 1964 by the first satisfactory highly-enriched uranium from the Lanzhou Gaseous Diffusion Plant. The product had been enriched to 90% and the achievement to this level in the short time since work began bears testimony to the degree of effort committed to the programme. There was great political pressure for China to demonstrate its prowess as an internal player – it was, after all, one of the five permanent members of the UN Security Council, the only one not yet having joined the atomic club.

While successful tests were under way with the DF-1, work was progressing rapidly on the development of a nuclear submarine, and while the H-6 bomber was being adapted and prepared for air-drop tests beginning in 1965 steps were made to achieve the first ground test.

The basic approach taken by China had been to acquire a bomb as quickly as possible. The realities of the threats China faced were apparent from the resolve of the United Nations to evict the North Korean aggressor from South Korea and that had been achieved with military force stopping only just short of the use of nuclear weapons by the United States. To obtain the necessary deterrent value of a nuclear-equipped country, China initially sought to simultaneously develop enriched uranium and plutonium, but in reality they barely managed to produce sufficient HEU for their first test.

The plutonium processing plants had been

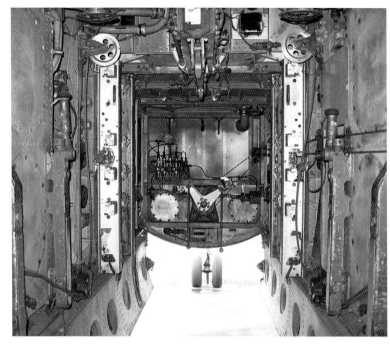

heavily compromised by the Russian withdrawal and it was not feasible to use plutonium in warhead design before it was finally produced in 1967. Consequently a significant shift now occurred as China redirected its efforts to achieve a nuclear capability with whatever means possible and they did not have the luxury of a twin-track approach as the Americans had opted for, and achieved, in the Manhattan Project a decade earlier.

Uranium mining had been the earliest industrial effort in the search to provide the raw material for processing and fabrication into a bomb and it was only through the Sino-Soviet agreement that work was conducted at a rapid

ABOVE The capacious bomb bay of the H-5, a copy and licence-built version of Russia's Ilyushin Il-28, procured on the very cusp of a falling-out between the two powers.
(David Baker)

LEFT In 1971 China tested a low-yield (8KT) device dropped from a Nanching Q-5 fighter-bomber of the type seen here. The aircraft was not designed for nuclear weapons but had been adapted by modifications to its single bomb bay.
(Via David Baker)

ABOVE The Dongfen 2 (DF-2) was China's first medium-range ballistic missile (MRBM) and had a range of 1,250km (776 miles) with a 20KT warhead. Used for China's first test of a nuclear-tipped rocket, it entered service in 1966 and was deployed for 20 years. *(IceUnshattered)*

pace, aided by access to an almost unlimited supply of workers. By the end of 1956 about 20,000 people were involved in prospecting for uranium, an intensity of effort reminiscent of the

BELOW The Sanjiang Missile Corporation developed the DF-11 to a requirement for a short-range ballistic missile (SRBM), a solid propellant round which is carried by a transporter-launcher and can be fired in less than 30 minutes from a decision to commit. The missile has a range of 300km (186 miles) and is nuclear-capable, which means that, like Russia's Scud series, it is usually equipped with conventional warheads. *(IceUnshattered)(Via David Baker)*

Gold Rush or the Klondike in North America during the 19th century. It was given the status of a national emergency and as such had call on untapped resources.

Two organisations had been set up in 1958 to work the mining operations and to develop processing technologies: the Hengyang Uranium Mining and Hydro-metallurgy Design and Research Academy at Hunan, and the Tongxian Uranium Mining and Hydro-metallurgy Institute. From these two organisations flowed the necessary resources and planning to scour the country for ore. Hengyang also served as a training facility for prospecting engineers, and processing plants were set up adjacent to the mines to simplify logistical challenges.

The first eight fully operational uranium mines were open by 1962–63, but in this somewhat crude effort – achieved more by brute force than subtlety – the Chinese were able to provide 150 tonnes of uranium concentrate quickly enough to shorten the preparation time for the first detonation by at least a year. In due course China would open 26 fully developed uranium mines, which expanded production proportionately.

Placed in perspective, of world uranium deposits, China has a mere 1.2% and is almost at the bottom of the world table, Australia possessing the most at almost 23% of world quantities, followed by Kazakhstan (15%), Russia (10%), Canada (8%), South Africa (8%) and the US (6%), among others. Extraction is not proportionate to quantity and the mining operations in China proved to be among the toughest in the world. But mining is only the

RIGHT With a range of up to 12,000km (7,450 miles), the DF-5 emerged from the China Academy of Launch Technology as a two-stage liquid propellant ICBM that began to enter operational deployment in the early 1980s. The first stage of the DF-5B shown here is more powerful than that for the DF-5 or -5A and the missile now has a range of 15,000km (9,320 miles). *(IceUnshattered)*

CENTRE The second stage of the DF-5B, which is capable of supporting between three and eight MIRV warheads or a single 4–5MT device while some tests have been conducted with a ten-MIRV configuration. *(IceUnshattered)*

first stage and China was challenged by the necessary enrichment processes required to turn it into high-grade material.

Using techniques similar to those in the US, the UK and France, the uranium ore was separated into concentrated U_3O_8 yellow cake at mills located adjacent to the mines and the ore then converted into uranium tetrafluoride (UF_4) at other locations. UF_4 was processed into UF_6 and the gaseous phase used on gaseous diffusion enrichment plants to concentrate the isolated U-235 as HEU. Initially, the Russians had agreed to supply China with UF_6 but with their withdrawal indigenous production was the only way to move toward a bomb.

To accomplish this, China built the Nuclear Fuel Component Plant at Baotou to specialise in UF_6 production, as well as other products, which included uranium fuel rods for the submarine nuclear reactor at the Subei complex. It was also responsible for the lithium-6 deuteride and tritium. Production of UF_6 was under the Institute of Atomic Energy Science near Beijing and later at the Subei

RIGHT The DF-16 has replaced the DF-15, which had been produced by the China Aerospace Science and Technology Corporation (CASC) as a single-stage SRBM with a range of 600km (373 miles). The DF-16 has a range of up to 1,600km (994 miles) and is now in the MRBM category with a high lofting trajectory that makes it more difficult to intercept as it spears toward its target. *(IceUnshattered)*

complex, built as part of the Jiuquan facilities. An additional plant was built later in Sichuan. To understand and develop the principles of gaseous diffusion, which the Russians had refused to show the Chinese, a special laboratory was set up at the Institute of Atomic Energy in Tuoli for evaluating the best way to achieve enrichment.

The template for China's gaseous diffusion plant was the Oak Ridge facility in the United States, about which the Chinese had considerable information and some personal knowledge, but using Soviet designs for the engineering layout. The selected site was situated 25km (15 miles) north-east of Lanzhou on the Yellow River and was approved on 31 May 1958. While the Russians had agreed to supply the design for their Verkhniy-Neyvinskiy gaseous diffusion plant in the Ural Mountains, they did not allow the Chinese to have any drawings of their own selected location, providing only design configuration drawings rather than layout plans.

This delay added two years to the development of China's bomb but in 1963 they sent the first uranium hexafluoride to Lanzhou for test runs of the enrichment cascade. Development was relatively slow, but by the early 1970s the facility would be producing 150–330kg (330–826lb) of weapons-grade E-235 each year. From this success the Chinese advanced to carry out research on the more difficult process of centrifuges for enriching uranium and they set up two research institutes, testing the first separator prototype

in 1981. Despite this apparent progress, China continued to use the gaseous diffusion method right into the 1990s.

Success

But all of that was in the future. China's first atom bomb was detonated at 07:00 UTC on 16 October 1964 at a place called Lop Nur, about 70km (43 miles) from Lop Nur dry lake. It was situated on top of a tower 102m (335ft) tall and delivered a yield of 22KT. The bomb was a uranium-235 implosion fission device that ignited concern across south-east Asia, not least from the Taiwanese who demanded an immediate military response from America, which was at the time already getting heavily involved with military action in a region to the south. But, political fallout notwithstanding, the bomb was a technical triumph for a country that had sought equality at the 'top table' and had achieved that despite extraordinary challenges.

The bomb weighed 1,550kg (3,418lb) and, with a cynical snub to their former ally, was nicknamed 596 for the month and the year (June 1959) that Russia formally withdrew all support. Electing to use the implosion design, rather than the gun-type assembly that was technically easier, the method required less fissile material. The detonation was witnessed by more than 5,000 technical and security personnel and was conducted with 21 effects experiments, including nine military units arranged around the test area. Placed at various distances up to 200m (656ft) from ground zero and a maximum 3,000m (9,840ft), these included tanks, aircraft, artillery weapons, the superstructures of naval vessels, communications equipment, animals, oils, medicines and food.

While 596 had been a uranium fission bomb, the search for a plutonium supply involved ground broken for the Jiuquan Atomic Energy Complex in Subei county in February 1960, and this quickly became a critical hub in the whole operation for plutonium production, a time when the Russians were pulling out. The reactor was not completed until 1967 followed by the reprocessing plant at Jiuquan in late 1970. The complex includes a reprocessing plant, a nuclear fuel processing plant and

BELOW Carrying a single 500KT warhead, the DF-21 was produced by the China Changfeng Mechanics and Electronics Technology Academy as a two-stage, solid-propellant missile with a range of 1,770km (1,100 miles). The missile has been the basis for development of the JL-1 submarine-based missile. Around 100 are believed to be in service.
(IceUnshattered)

RIGHT The DF-21D variant is an anti-ship missile capable of taking out a carrier task group, in which role it has a range of 1,450km (900 miles) and can threaten any surface vessels or fleet within strike-range of the Chinese mainland.
(IceUnshattered)

CENTRE Developed by the CASC, the DF-26 is a solid-propellant IRBM with a range of up to 4,000km (2,485 miles), road-mobile and capable of carrying conventional or nuclear warheads. It is believed to have entered service in 2015 and is conveyed on a transporter-launcher.
(IceUnshattered)

a manufacturing and assembly workshop, employing more than 10,000 workers.

The method initially transferred from the USSR was believed to be the difficult and complicated sodium uranyl acetate precipitation process, time-consuming and difficult to achieve. Moreover, the recovery rate was low – the uranium could not be recovered and great quantities of liquid waste were produced. The Chinese quickly adopted an alternative method, the plutonium-uranium extraction technique that is known as PUREX in the United States. The acronym stands for plutonium-uranium redox extraction, which has become the standard aqueous reprocessing technique for recovering uranium and plutonium from spent fuel and is based on the liquid-liquid extraction ion-exchange process.

This technique has been challenged in America as being unsafe, particularly after the Hanford site was found to have produced 'copious volumes of liquid wastes', according to a government report in 1992 which estimated that harmful quantities of radioactive iodine had been released into the atmosphere. However, concerns of this nature were not priorities in the

RIGHT Employing an astro-navigation system using satellites, the DF-31 ICBM entered service in 2006. With a range of up to 11,200km (6,960 miles), the DF-31A carries a single 1MT warhead while the DF-31B can launch with a cluster of three to five MIRVs of 20–150KT yield each.
(IceUnshattered)

China of the Cold War era and a pilot plant was built in only three years, material being placed in the pilot plant successfully for the first time on 4 September 1968, followed by activation of the main plant on 18 April 1970. But there were problems. Poor design and construction imperfections haunted operations and several flaws made the processing intermittent.

Continuous modification, learning and reapplication of proven methods of purity and quality allowed full production to be achieved by the 1980s and a new plant was built during the 1990s. The Nuclear Fuel Processing Plant took enriched uranium hexafluoride from the Lanzhou facility and converted that into uranium tetrafluoride for additional processing, shaped into bomb cores for further processing into metal at the Nuclear Component Manufacturing Plant. After the plutonium was separated into pure plutonium metal it was worked into fissile components, with final assembly at the Assembly Workshop of the Nuclear Component Manufacturing Plant.

It took China a mere eight months from its first atom bomb test to prepare and demonstrate a weaponised device dropped from an aircraft in flight. On 14 May 1965 the Chinese dropped a 35KT fission device from its H-6 bomber and repeated that on 9 May 1966 with its third nuclear weapon test, a 100KT bomb also dropped by the H-6. This boosted-fission weapon contained uranium-235 and lithium-6 thermonuclear fuel as a first step toward a true fusion bomb and the concept was tested for a second time on 28 December 1966 in a tower test where a yield of 300–500KT was delivered.

This second pre-thermonuclear test included an effects package of 81 experimental items including animals, aircraft hangars, missile launch shafts, underground aircraft hangars, underground railroads, dams, tunnels and simulated fortifications together with various items of weaponry. The site was hardened with reinforced cement and stone blocks out to a radius of 230m (755ft). The test was directed by Nie Rongzhen and 28 military units involving 6,400 personnel.

With several lines of development converging rapidly toward full nuclearisation of its defence capabilities, on 27 October 1966 China fired a modified DF-3 carrying a 12KT fission bomb a distance of 894km (555 miles), the warhead detonating at the pre-set altitude of 569m (1,870ft). This road-mobile weapon had a maximum range of 1,250km (776 miles) and was deployed to north-east China, targeting cities and US military bases in Japan.

Not as effective as it might have been, the DF-2 series used a combination of non-storable alcohol and liquid oxygen propellants that required several hours of preparation for launch. As it was, it served its purpose as a strident reminder that China was prepared to defend itself with rocket-boosted nuclear weapons, and from the combination of air and missile forces it was already a power that could inflict unacceptable damage on a neighbouring aggressor.

No sooner had the peak of activity on the fission bomb been reached than resources were

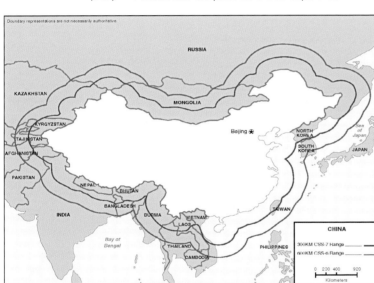

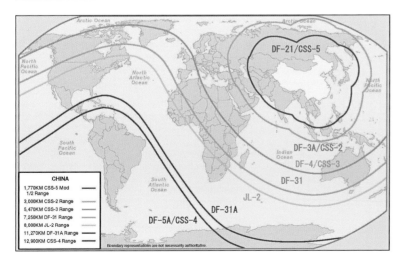

directed toward thermonuclear capability with megaton-class yield dropped by air or fired by missile. Great progress was made and China's first thermonuclear weapon was dropped from an H-6 bomber on 17 June 1967, detonating at an altitude of 2,960m (9,710ft) with a yield of 3.3MT. Some 53 effects items were emplaced at various distances and a solid propellant rocket was launched directly into the mushroom cloud to collect radioactive samples. The bomb was a two-stage device using U-235, U-238 and lithium-6 plus deuterium.

A second test was attempted on 24 December 1967 with U-238 and lithium-6 but it was unsuccessful and fizzled at 10–12KT. The first test of a plutonium bomb took place with the next test on 27 December 1968, which was dropped by an H-5 bomber, a Chinese copy of the Ilyushin Il-28 medium bomber. The first underground test took place on 23 September 1969, a 20KT device, and this was followed by a succession of further underground tests demonstrating various combinations including a low-yield design in November 1971 with 2kg (4.4lb) of boosted plutonium primary with 0.5kg (1.1lb) of Oralloy (named after Oak Ridge Alloy) or highly-enriched uranium by special process. The following test was a low-yield 8KT device dropped from a Q-5 (Qian-5), a supersonic attack aircraft with a single internal bomb bay designed for a nuclear weapon.

Chinese tests

The first test of a Chinese nuclear device on 16 October 1964 at the now famous Lop Nur facility was followed on 14 May 1965 by an air-dropped weapon with a yield of 35KT. China's third test was their first use of a thermonuclear primary and had a yield of 250KT when detonated on 9 May 1966 but the fourth test on 27 October 1966 was a 12KT warhead delivered by the CSS-1 Dong Feng 3 MRBM across a range of 894km. Confirmation of a two-stage boosted fission came on the fifth test, 28 December 1966.

In keeping with its methodical and paced programme, China detonated its first thermonuclear device on 17 June 1967, an air-drop weapon with a yield of 3.3MT. Nuclear testing continued at a persistent pace with several tests each year but at a level commensurate with a steady and consistent expansion of capabilities designed to provide a nuclear deterrent at land, at sea and in the air.

By the date of the last test on 29 July 1996 China had conducted 47 tests with 48 devices, delivering 4.5% of all nuclear test yields worldwide by all nuclear powers.

The Chinese stockpile

Of all the known nuclear powers, China has been one of the more restrained and composed in response to potential adversaries on their doorstep and farther away. In proportion to their military strength, China has quantitatively fewer nuclear delivery systems than would be considered proportionate to its overall military and economic power.

BELOW The *Xia* nuclear submarine entered service in the late 1980s as China's first ballistic missile boat, with a capacity for 12 JL-1A SLBMs. Designated Type 902, it is being replaced by the Type 904. *(CNS)*

The development of China's nuclear land, sea and air delivery systems evolved gradually since initial deployments in the 1970s. The disruption caused by Mao Zedong's Cultural Revolution of 1966–76 had little effect on the development of the strategic military programmes, or its assets, and the work progressed largely unhindered by the convulsions of change that swept over China, in which it is believed several million people were killed or persecuted.

There is still a range of secrecy regarding the precise numbers of nuclear weapons and delivery systems but there is clear evidence for the internationally accepted values quoted here. Of land-based strategic missiles, China has 62 ICBMs of which ten are the surface-based DF-4 with a range of 5,470km (3,400 miles), 20 are the silo-based DF-5 series with a range of 12,000km (7,450 miles), eight are rail/road-mobile DF-31 missiles with a range of 8,000km (4,970 miles) and 24 are the DF-31A with a range of 14,000km (8,700 miles).

Initial attempts by the Chinese to develop a sea-launched missile consisted of a navalised version of the Russian Scud A, but the liquid propellant combination required the submarine to surface in calm sea conditions before preparation of the missile for firing. Acquired in early 1960 just before relations with Russia

soured, the renamed 1060 missile was only a temporary solution to the requirement and not for another 25 years would China deploy its first nuclear-powered ballistic missile submarine.

With cautious but protracted development, China only succeeded in testing a pressurised water-reactor in August 1970 and the reactor in a *Han*-class attack submarine achieved criticality in June 1971, preceding sea trials from August that year to the end of 1972. The first *Han*-class SSN was turned over to the Navy in August 1974, and these submarines were used to develop and refine the technology required for missile-carrying boats. China's first underwater launch of a missile on test took place on 12 October 1983 when a JL-1, China's first underwater missile, was launched from a Soviet *Golf*-class submarine that had been procured from Russia before 1960.

The first missile boats went into service in 1986 as the *Xia*-class and at first it was believed that two were built, each equipped with 12 missile tubes, but no evidence has surfaced to substantiate this and only one is known to have taken to sea. Equipped with the JL-1, the first test launch of this missile took place on 30 April 1982. With a range of 1,770km (1,100 miles), it carried a 250–500KT warhead and was superseded by the JL-1A with a range of 2,500km (1,550 miles).

Currently China has four nuclear submarines with a total 48 Jl-2 SLBMs and each carries either a single 1–2MT warhead or up to three or four 20, 90 or 150KT MIRVs, together with decoys and penetration aids. This is comparable to the force level the UK Royal Navy will have when re-equipped with the *Dreadnought*-class and Trident II D5. Each JL-2 has a maximum range of 7,200km (4,475 miles). The type was launched for the first time in 2001.

China has 16 dual-capable DF-26 IRBMs, which became operational in early 2015 and have a range of 3,000–4,000km (1,900–2,500 miles) with either a conventional or nuclear warhead. It has deployed 146 DF-21 MRBMs from which the DF-21 was developed. It has a range of 1,770km (1,100 miles) for the most powerful derivative variant, of which 80 are nuclear-capable and 66 carry conventional warheads.

The DF-11 is a land-mobile, short-range

BELOW A comparative scale representation of the JL-1 and JL-2 SLBMs, the latter destined for the Type 904 submarine. A navalised version of the DF-31, the JL-2 has a range of more than 7,500km (4,660 miles) carrying a single 250KT–1MT warhead or up to four MIRVs of 90KT yield. *(US Office of Naval Intelligence)*

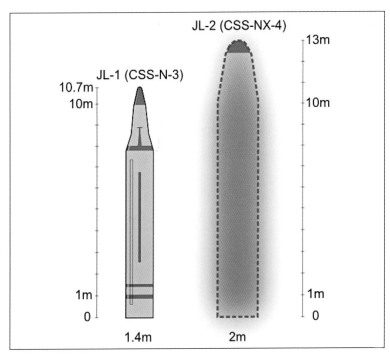

ballistic missile (SRBM) with a range of up to 350km (220 miles) or 600km (370 miles) for the DF-11A variant. Currently China has 108 deployed operationally. In addition it has 81 road-mobile DF-15Bs with a range of up to 800km (497 miles) with a 90KT warhead.

China completes its nuclear triad with an inventory of 60 H-6 and 60 H-6K bombers developed from the Tupolev Tu-16, each capable of carrying nuclear weapons. Considerable work is under way to improve China's nuclear capability but none of the current inventory is operationally deployed. While rumours abound over the number of nuclear weapons held by China there is little or no evidence to suggest that they have more than 300.

As asserted previously, China has a declared 'no first use' policy and its political and military statements have always insisted on the nuclear deterrent as just that and the country has projected a desire to shy away from brandishing them on the political stage. However, in the event of their use the perceived understanding of China's command and control structure, and the interfaces between elements of the infrastructure, leave no doubt that they would be employed with expediency and efficiency if all else failed and national survival was at stake.

ABOVE Currently China has four Type 904 submarines in service and is planning up to four more, each equipped with 12 JL-2 SLBMs and with a displacement of 11,000 tonnes submerged. *(US Office of Naval Intelligence)*

BELOW China's airborne deterrent is maintained by the venerable H-6 long-range bomber, introduced in 1968. The latest variant, the H-6K, employs turbofan engines. *(airliners.net)*

Chapter Seven

Other nuclear powers

Challenged by the nuclear powers, nation states across the globe sought atomic weapons in a proliferation of capabilities that today threatens regional security through the pre-emptive use of thermonuclear bombs on hostile neighbours.

OPPOSITE India's Agni-II ballistic missile, which has the potential to cover all targets in neighbouring Pakistan. *(David Baker)*

very foundation of the Indian nation in the 1940s, Prime Minister Jawaharlal Nehru inherited the national aspirations defined by Mahatma Gandhi and grew the technical capabilities of the country in science and technology. Eventually, especially after war with Pakistan, that resolve extended to building nuclear weapons.
(Via David Baker)

RIGHT India's Bhabha Atomic Research Centre and its nuclear reactor epitomise the deep resolve to embrace the atom both for energy and for strategic deterrence.
(BARC)

India

The drive for India to be a modern industrial, scientific and technological but independent nation grew in parallel with its determination to sever colonial links with Britain and to rely on the West for cooperation and dialogue in all those disciplines. Science had been an entrenched part of Indian culture for several centuries, long before British involvement in that country. A leading figure in India's nuclear programme would be Homi Jahinger Bhabha,

born in 1909, who had already gained an international reputation for nuclear physics before the outbreak of the Second World War.

Bhabha came from a successful family of industrialists. His uncle Dorabji Tata was already forging a heavy industry for India which would sustain it in the decades after independence was achieved on 15 August 1947. But Bhabha was to play an instrumental role in establishing pure research and in the development of a civil nuclear energy programme that would eventually lead to a nuclear weapons development effort.

A new way of interpreting the exchange mechanisms between energy and matter was created by Bhabha when he studied the relationship between matter and anti-matter. While the equitability of energy and matter was established by Einstein, Bhabha turned this on its head and posited that energy can be turned into matter through the annihilation of a positron and the release of energy, creating a gamma ray of lower energy. The work on interactions between electrons and positrons fed into a great international surge in the work on nuclear physics and this effect became known as Bhabha Scattering.

Under the premiership of Jawaharlal Nehru, Bhabha was to play a leading role in developing India's nuclear power programme, the acclaimed scientist now being seen as one of the more important people in developing India as a proud and independent country. With a background at Cambridge University, England,

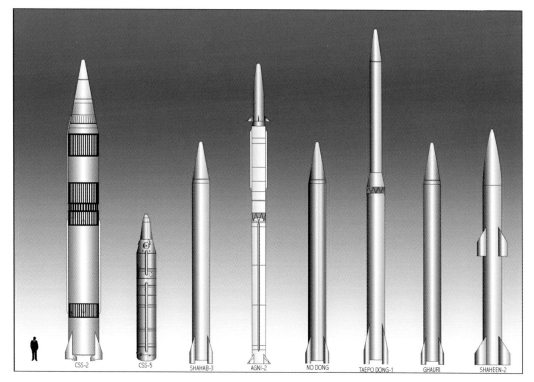

during the late 1920s and 1930s, he had been in precisely the right place to absorb the trends of the time and to have his ideas discussed. In 1946 he had been made chairman of the Atomic Energy Research Committee and two years later Nehru made him head of the Atomic Energy Commission with responsibility for government ownership of the country's uranium and thorium deposits.

This was an important appointment, as Bhabha had correctly identified the available natural resources of the country with the possibility of creating a peaceful nuclear energy industry. He reasoned that the country needed a three-stage programme due to the relative abundance of thorium – around 500,000 tons in the subcontinent (<2% of world total) – compared to uranium, about 45,000 tons (25% of global total). He reasoned that the programme must be based on the use of uranium to begin a reactor programme, the plutonium produced by that first generation being used to establish a second generation in which thorium would be converted into U-233 or depleted uranium into more plutonium with the advantage of a breeder capability. This second generation would produce more U-233 than required for producing energy.

This was the ideal way to configure India's

nuclear programme toward an autonomous capability for producing nuclear weapons, and while it had always been professed as a peaceful application Nehru had consistently grasped the reality of employing whatever scientific or technological advantages it could acquire to defend its interests. The connection between India's indigenous intellectual resources and the political disagreement with Pakistan is obvious.

The first stage in the three-phase programme funded construction of heavy water reactors from natural uranium to produce electricity and Pu-239 as a by-product. This was the classic employment of U-238 with deuterium oxide used as the moderator and coolant. To balance the meagre availability of its uranium resource, the government restricted power output to 13GW capacity to restrict the first-stage cycle. The total output of 4.8GW is handled by pressurised heavy water reactors and two boiling water reactors. The highest output (1.18GW) comes from the Rawatbhata plant, followed by Tarapur (1.4GW) and Kaiga (880MW). Another plant is being built at Banswara (4.2GW).

The second stage breeder reactors use a mixed oxide obtained from Pu-239 as the recovered reprocessed spent fuel from the first

stage, with natural uranium. Surplus plutonium can be stored for setting up more reactors and uranium from the first cycle can be used through multiple cycles in fast breeder reactors. After the Pu-239 has built up, thorium can be used as blanket material for transmuting into U-233 in the third stage.

The third stage thorium reactors operated as an advanced nuclear power system, envisaged as a thermal breeder, which, after its initial charging, would operate on natural thorium. The plan was to operate fast breeders until a capacity of 50GW had been achieved and then to move to stage three. This is a very long-term programme and will probably not be realised

until 2050, but the integration of the civil and weapons sides of nuclear development went in parallel to a closer degree than probably any other country to date.

Homi Bhabha began lobbying for a weapons programme in the early 1960s, seeing it as a means of strengthening India's national defence, and used the subtle argument of nuclear detonations in the service of humanity by adopting their use in applications proposed in the United States by Project Ploughshare. Bhabha was unable to gain full support for India's nuclear weapons programme until the Chinese detonated their own device in October 1964. Thus began a political chain reaction that would result in Pakistan wanting a bomb as well. However, for some time the development of nuclear weapons slowed, albeit being reinvigorated when Indira Gandhi became Prime Minister in 1967, with Homi Sethna playing a key role in working out the development of weapons-grade plutonium, preferred over uranium.

In January 1969, based on a survey of Russia's Dubna plutonium-fuelled fast reactor, India's plutonium production plants were formally approved. But it was after the war with Pakistan in 1971 that work accelerated, despite a certain ambivalence about nuclear weapons and their use. Work had begun on the first top-secret plant, known as Purnima, in January 1969, and on 7 September 1972 Gandhi specifically instructed the Bhabha Atomic Research Centre (BARC) to design, develop and test an atomic bomb. The programme was managed by Raja Ramanna, director of BARC.

India has refused to sign the Nuclear Non-Proliferation Treaty on the basis that it favours the major nuclear powers and was unconstrained in efforts which those powers frequently attempted to subvert. The designer of the bomb was P.K. Lyenger, the deputy manager, assisted by chief metallurgist R. Chidambaram and N.S. Venkatesan, who designed high explosives to produce implosion. The actual materials and the detonation system were produced by W.D. Patwardhan from the High Energy Materials Research Laboratory.

Much of the general configuration of the bomb was taken from the design of the US Fat Man dropped by the Americans on Nagasaki. The 6kg (13.23lb) of plutonium came from the

PRESIDENT

BARC facility's Canadian-Indian Reactor Uranium System (CIRUS) reactor in Trobay near Mumbai, supplied by Canada in 1954. CIRUS had been approved under the explicit understanding that it would be used only for peaceful purposes, as was the heavy water provided by the United States. The polonium-beryllium neutron initiator was code named Flower.

Assembled at Tromboy and known colloquially as the *Smiling Buddha*, India's first implosion device had an hexagonal cross-section of 1.25m (4.1ft) with a weight of 1,400kg (3,087lb) and mounted on a metal tripod placed in a shaft 107m (351ft) beneath the surface of the Army's Pokhran test range in the Thar Desert. Detonated on 18 May 1974, it had a yield of approximately 4–6KT, although the scientists claimed it was 13KT.

World reaction was spontaneous and disapproving, despite India asserting that it was a peaceful use of atomic power much as the Americans had proposed with the Ploughshare concept, but that hid nothing from a world concerned about proliferation. The detonation resulted in the formation of the Nuclear Suppliers Group (NSG), which sought to control the export of materials and technology used to assemble nuclear devices. It first met in November 1975 with seven members, which by 2016 had grown to 48 countries pledged to constrain proliferation.

Formally known as the Pokhran-I test, *Smiling Buddha* was followed by Pokhran-II on 11–13 May 1998, a series of five detonations achieved despite the efforts of the NSG, which placed an embargo on support for both India and Pakistan. In the 24 years between these two test phases, work had continued on a fusion project for a hydrogen bomb developed under M. Srinivasan. Tests were to have resumed in 1995, but after spy satellites picked up information about preparations extensive pressure from US President Bill Clinton on India's Prime Minister Narashimha Rao brought about their cancellation, pressure boosted by a vitriolic outburst from Pakistan's Benazir Bhutto.

ABOVE The development of nuclear power preceded the real commitment to a nuclear weapons industry and attention focused at first on building reactors for energy, such as this facility at Kudankulum. *(BARC)*

BELOW An anachronism defined by historical India, a fort near the Pokhran test site for India's first nuclear weapons connects across the centuries to define the tensions still present with contested neighbours. *(David Baker)*

RIGHT A poor-quality image shows the dust raised by two simultaneous nuclear detonations during the Shakti nuclear weapons test series held at Pokhran in 1998. *(Via David Baker)*

A major intelligence operation was mounted by the West against these two countries, but for geographic reasons it was easier for Pakistan to hide its own test preparations. When the Bharatiya Janata Party came to power in India in 1998 there was increased pressure from within for a set of tests, but secrecy was of the utmost and most members of the government knew nothing of these preparations. Special test shafts were dug under camouflage nets, cables for recording data were covered with sand, native vegetation was used to conceal small working huts and recording stations and concealment of transport reached unprecedented heights.

There were to be five detonations, coded Shakti, with the devices from BARC delivered to the Jaisalmer Army Base by Antonov AN-32s of the Indian Air Force. Again, a highly secret operation swung into action to maintain the tightest veil around these preparations. The five detonations were organised at two sites, with two in one shaft, and all devices at the separate locations fired simultaneously. Shakti I was a thermonuclear device with a yield of 45KT but a design yield of 200KT when deployed, with Shakti II a plutonium implosion device with a 15KT yield and specifically designed for air carriage or missile application. This was an improved design over the Smiling Buddha and had been developed using computer simulations.

Shakti III was of a linear implosion design utilising non-weapon-grade plutonium with a yield of 0.3KT, while Shakti IV was an experimental shot delivering 0.5KT and Shakti V was a similar detonation yielding 0.2KT. The last two were too low in yield to record seismic profiles for analysis. Some reports indicate that a sixth shot was planned but failed to detonate. The Pokhran-II detonations brought universal acclaim in India, the Bombay Stock Exchange soaring when the news broke, but international opinion was outraged and sanctions were imposed.

Today India has a stockpile of around 110–120 weapons with intelligence assessments concluding that the country has produced about 540kg (1,190lb) of weapons-grade plutonium. This is sufficient for up to 180 nuclear weapons but not all of it has been converted for that purpose. The primary reactor at Dhruva near Mumbai is being joined by a second facility near Visakhapatnam on the east coast and a fast breeder reactor has been built south of the newly renamed Indira Gandhi Centre of Atomic Research.

India has a triad of nuclear delivery platforms for air, land and sea forces, the air element consisting of 48 Jaguar IS/IB and Mirage 2000H aircraft with a range of 1,600/1,850km (994/1,150 miles) capable of deep strikes into China or Pakistan. The French Mirage has served a nuclear strike role with French air units for several decades and the Jaguar was nuclear-capable from the outset. Some reports suggest that their 90 MiG-27 Floggers have a nuclear strike mission and this may evolve more effectively as the additional plutonium from the second reactor plant increases the quantity of material available for bombs. Similarly, some of their Sukhoi Su-30MKIs (of which they currently have more than 200) may soon have a nuclear role and, with a range of 1,000km (620 miles), the Nirbhay cruise missile will add robust improvements to its attack potential.

Land-based missiles include the Agni series of which four types are nuclear capable, accounting for an assigned inventory of 56 warheads as of 2017. First deployed in 2003, the 30 SS-250 Prithvi-2 short-range missiles have a range of 250km (155 miles) and a yield of 12KT. The Agni-I was deployed from 2007 with a range of 700km (435 miles), the Agni-II from 2011 with a range of more than 2,000km (1,243 miles), the Agni-III from 2014 with a range of more than 3,200km (1,988 miles), the

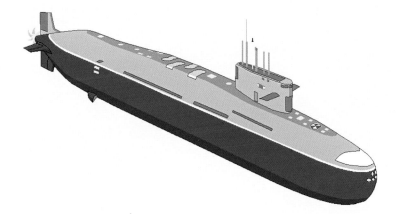

Agni-IV from 2016 with a range of 3,500km (2,175 miles) and the Agni-V from 2017 with a range of 5,200km (3,230 miles). All these Agni missiles carry 40KT warheads.

Development of these missiles has carried the deterrent to road-mobile and rail-mobile deployments, with the Agni-V being in a sealed container that allows it to be fired within minutes of arriving at its launch site. The extended range also allows it to be deployed further from China or Pakistan to raise its survival potential.

India has been developing two naval nuclear weapons systems, a nuclear-powered ballistic missile submarine and a ship-launched ballistic missile. *Arihant*, India's first SSBN (Ship Submersible Ballistic Nuclear), went on trial in 2014 equipped with 12 launch tubes for the 700km (435 mile) range K-15 Sagarika missile. A second submarine is under final fitting and a third is under construction. The limited range

ABOVE Declared operational in February 2016, India's Arihant submarine marks a new and more vigorous commitment to sea-based nuclear weapons platforms, each submarine capable of carrying 12 K-15 missiles with a range of up to 1,900km (1,180 miles). *(Gagan)*

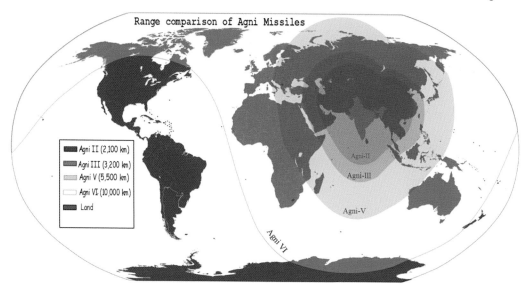

Range comparison of Agni Missiles

- Agni II (2,100 km)
- Agni III (3,200 km)
- Agni V (5,500 km)
- Agni VI (10,000 km)
- Land

LEFT India's expanding missile programme has provided a progressive evolution toward increasingly more powerful rocket delivery systems, as shown on this map of range capabilities for selected types. *(DoD)*

of the K-15 is a serious impediment to India's boast of a fully effective nuclear triad and the Dhanish sea-based missile, with a range of 400km (249 miles), does little to improve the picture. It is designed to be carried on the stern of converted Sukanya patrol boats.

Pakistan

The development of a nuclear weapons programme for Pakistan is identified under Project-706, which was that country's first fission device using uranium, which eventually matured into a reactor-grade plutonium programme for weapons-grade material in the 1980s. It is frequently quoted as Pakistan's nuclear weapons programme from its origin in the late 1940s to the present, but that is not so. Project-706 originated in the early 1970s and was disbanded when the first cold test of an atomic device was achieved on 11 March 1983.

The origin of Pakistan's push for nuclear weapons lies in a complex geopolitical history of that country's origin as an Islamic republic in what is one of the most important strategic locations in South Asia, bordered by India, Afghanistan and China and with a 1,050km (650 mile) coastline washed by the Arabian Sea and the Gulf of Oman. Created in 1947 as, still, the world's only country set up as an Islamic state, it attracted a wide range of scientists from India, with the active encouragement from the then Prime Minister Liaqat Ali Khan.

Support for a domestic nuclear energy programme came from a wide range of sources, including the nuclear physicist Mark Oliphant, who wrote a letter to Pakistan's Muhammad Ali Jinnah suggesting such a programme. By the early 1950s Rafi Muhammad Chaudhry had been encouraged to come to Pakistan and set up a high-tension laboratory, which he did in 1952. Encouraged by President Eisenhower's Atoms for Peace plan, Pakistan was an early signatory while absolving itself of any interest in developing nuclear weapons.

This situation prevailed during a period of expansion of Pakistan's civil nuclear power programme but that changed after the war with India in 1965. There was some pressure from the military for Pakistan to develop its own nuclear weapons and this intensified when India expanded its own nuclear programme and moved toward a weapons capability. But no programme was instituted. Conflict and unrest fed into a general feeling of isolation, exacerbated by a 13-day war with India in 1971, when Pakistan lost a large part of its territory, a high percentage of its military and several millions of its population to the newly created state of Bangladesh.

Nobel Laureate Abdus Salam started work on an atomic bomb in December 1972 and recruited two junior physicists from the International Centre for Theoretical Physics to work under Munir Ahmad Khan on plans for a major development programme that were

BELOW Keen to embrace all facets of modern technology and to provide Pakistan with a nuclear deterrent, Prime Minister Ali Khan meets the President and faculty of the Massachusetts Institute of Technology in 1950.
(Via David Baker)

BELOW RIGHT Sir Mark Oliphant played a leading role in encouraging Pakistan to develop a nuclear energy programme, emphasising the opportunities that were available through the Atoms for Peace programme, which sought to assist developing countries while restraining their acquisition of nuclear weapons.
(Via David Baker)

presented to the Pakistan Atomic Energy Commission. Soon a Theoretical Physics Group was formed to carry out original work on fast neutron reactions, neutron diffraction, simultaneity, hydrodynamics and the essential requirements for building a working bomb, and then how to make that bomb work. Special departments and working groups were set up to carry out theoretical calculations and to prepare the way for a formal programme.

Work on Project-706 got under way in March 1974 at the Pinstech Institute, with Salam and Ahmad playing a key role. Funding for the project, an estimated $450 million, was raised quietly by Saudi Arabia and Libya and a veil of secrecy descended over the entire project. It was decided to adopt the implosion design rather than a gun-type device, and several teams were sent out to locate resources of uranium and natural plutonium ores. Thus were set in train the steps necessary for converting U-235 into Pu-239, production of which was under the Nuclear Physics Group (NPG), formed in 1967.

As work progressed, on 22 May 1974 Pakistan received immediate word of India's detonation of a fission bomb and viewed it as a provocative act designed to position that country as the predominant power in the region. Thus energised, work on Pakistan's bomb received high-level priority and the country's armed services established their own technical boards to apply their own research capabilities to the issue. Noted military engineer Brigadier Zahid Ali Akbar Khan was placed in charge as head of the facilities' construction and played a role not dissimilar to that of General Groves in America's Manhattan Project.

It was about this time that the plutonium production work was placed under Munir Ahmad Khan, along with uranium production and the entire exploration, mining and processing operation. But there were concerns from several foreign powers regarding the suspected development of a Pakistani bomb, which some hailed as an 'Islamic weapon of mass destruction'. Both Russia and the United States engaged in an intensive intelligence-gathering exercise, and Pakistan's own intelligence organisations were busy arresting and hunting down spies and espionage agents from these and other Middle East states.

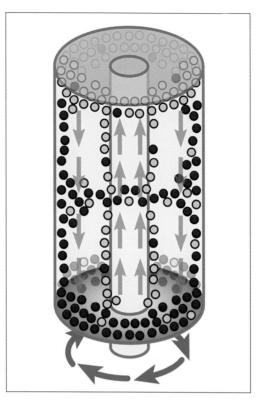

In 1981 Israel conducted an air strike on Iraq's own nuclear weapons development programme, destroying a reactor and setting back that effort by at least ten years. Sensitive to a similar strike against its own emerging capability, Pakistan stiffened its air defences and successfully defended against an intrusive strike by locking on to incoming aircraft just short of the border. Israel desisted from pressing home its intentions to destroy Pakistan's nuclear facilities.

A remote region in Sargodha was chosen as the site for initial testing of Pakistan's highly secret atomic bomb. Twenty-four 'cold' tests were carried out to simulate various stages in the detonation of a bomb but without an explosive result. These tests were conducted between 11 March 1983 and 1990 with overall supervision under the control of Ishfaq Ahmed. But this was far short of the potential for an actual bomb and merely confirmed that the scientists and engineers could induce a chain reaction. These tests were verification of concepts and not stages on the way to a detonation.

With Project-706 having provided the launch of the uranium programme, development work for a bomb with production of highly-enriched uranium proving difficult, assertions by A.Q. Khan that he could achieve that with the gas

LEFT In Pakistan, Dr A.Q. Khan organised a new concept of centrifuges adopting similar principles to the Zippe-type gas centrifuges in which U-238 is in dark blue and U-235 in lighter blue. An even mixture of the two is released into the centre of the centrifuge where inverse centripetal forces move the U-238 to the outer edge. By heating the bottom, convective currents concentrate at the top. By the early 1980s there were up to 2,900 gas centrifuges in operation. *(Via David Baker)*

ABOVE Pakistan's Khara Desert area selected for its nuclear weapons test programme. *(NASA)*

BELOW The Chaghi Monument in Islamabad, a tribute to the successful efforts of Pakistan's nuclear scientists and engineers with a successful test on 28 May 1998. *(David Baker)*

centrifuge system brought opposition from colleagues. He persisted, and succeeded, with Pakistan's first successful underground nuclear detonation on 28 May 1998 with a yield of 32KT. It was a boosted fission device and four more detonations took place at the same location that day, each with a yield of 1KT.

The Chagai-1 test was conducted in an isolated region of Pakistan located in the Ras Koh Hills, a site selected as early as 1978 for its dryness, geologically stable structures and lack of human population. International reaction was quick and condemnatory, accusing Pakistan of violating agreements on non-proliferation and of creating tensions within the region. It was nevertheless followed by Chagai-2, an underground test with a yield of 15KT, a miniaturised boosted fission device.

This second test was of a plutonium bomb capable of being deployed with missiles, aircraft and ships, a preparation for the production of bombs incorporating tritium boosters. It is this line of development that sustained a commitment to plutonium-based weapons and for which some expansion of capability has been observed by intelligence sources within the last decade. Political pressure to stop work on this has been just as robust as regional challenges and the potential threat from adjacent states.

The development of an indigenous plutonium capability began in the early 1970s and sustained this evolution alongside the uranium programme. But unlike the uranium work, plutonium production was a wholly national affair and the Pakistan Atomic Energy Commission (PAEC) – which had been formed in 1956 – led the work under its chairman Munir Ahmad Khan. This work brought international opposition and an embargo that slowed development on the construction of an electromagnetic isotope separation process.

By the end of the decade the PAEC was able to start planning plutonium production from its 40–50MW Khushab complex at Joharabad but not until 1986 did construction work begin. This heavy-water reactor is capable of delivering 8–10kg (17.6–22lb) of plutonium each year and this production rate is expanding proportionate to additional facilities at the complex. It can also produce H^3 or Li^6 but this is not necessary for current state-of-the-art weapons as they use

RIGHT With an emphasis on road-mobile capabilities, Pakistan shows off its medium- and intermediate-range missiles. *(Via David Baker)*

CENTRE The Babur ground-hugging cruise missile has been developed by Pakistan in response to India's alleged interest in a US Patriot anti-ballistic missile system. The transporter-launcher can carry four cruise missiles, seen here in Karachi. *(Via David Baker)*

[6]Li. Work began on a significant expansion of the plutonium facility in 2006 and the estimated output from this reactor will be sufficient to produce 40–50 nuclear warheads each year.

The stockpile of Pakistan's nuclear deterrent is hard to identify precisely but a wide range of estimates in the public domain focus on 100–140 warheads. Production of HEU is certainly high enough to provide many more warheads than this, given that in a modern warhead a critical mass of 90% enriched uranium is about 52kg (115lb), or 8–10kg (17.6–22lb) of plutonium. However, Pakistan's implosion-type bombs would require only 25kg through the use of neutron reflectors and a tamper of beryllium. With a similar refinement only 2–4kg (4.4–8.8lb) of plutonium is required for the same amount of HEU required to produce the same yield.

Refinements to stockpile estimates can be had from known technologies and installations. For instance, the Kahuta HEU production facility has 10,000–20,000 centrifuges which could theoretically support almost 100kg (220lb) of HEU per annum since 1986, or 1,500kg (3,307lb) by 2005 which would have been sufficient for 75 warheads or bombs. Overall analysis, combined with known readings of the magnitude of the Chagai-1 and -2 tests indicates that Pakistan has two classes of

RIGHT The Babur has deployable flying control surfaces with a top speed of Mach 0.8 and a range of up to 700km (435 miles) carrying conventional or nuclear warheads. With terrain-contour-matching radar and a navigation system compatible with Russia's Glonass or China's Bedu satellite navigation system, Babur was deployed from 2009. *(Via David Baker)*

bomb, one with yields of 20–25KT and a second group of 150–500KT, the latter divided between two separate classes of launch vehicle.

Unlike other minor nuclear states, Pakistan entered into a study of how it might survive a nuclear exchange and preserve a sufficient strike force to conduct a nuclear response after riding out the initial repercussions of a surprise attack. To achieve this would be to consolidate the deterrent effect, eliminating the desirability of an aggressive surprise attack on the basis that such a strike would not remove the consequences of retribution to the aggressor state. This second-strike capability has driven a sustained expansion of Pakistan's nuclear forces and has prompted the government to commit unprecedented resources to hardened storage and launch facilities together with protected bunkers for government command and control not only of the armed forces but of the administrative and general population too.

Since 1971 Pakistan has had a no-first-use policy, but the ability for a second strike could be interpreted by its potential enemy, India, as a threat to peace in the region due to the leverage it would have in sustaining a nuclear exchange and prevailing. Pakistan sees India as its greatest threat to peace and that has driven the country's geopolitical map for several decades.

The country has set four specific circumstances in which it might go nuclear: spatial threshold where India occupies large parts of the country; military threshold where a large part of the defence forces are destroyed; economic threshold where the Indian Navy blockades Pakistan from sea trade; and political threshold where internal destabilisation is clearly and unambiguously the responsibility of another state, implying India.

For their part, potential enemies have highlighted the role played by some scientists in Pakistan in the nuclear black market, Abdul Qadeer Khan having been arrested in 2004 for selling nuclear weapons technology and information to Iran, North Korea and Libya. Moreover, concern has been expressed over mixed messages. In 1998 the country's foreign secretary, Shamshad Ahmad, proclaimed Pakistan's right to use nuclear weapons in the event that India attempted to prise apart a factional section of the country, as happened in the Bangladeshi liberation war of 1971.

Pakistan's stockpile today is about 140 nuclear warheads but that is increasing at a rate that has outstripped projections made by US intelligence. Approximately 36 aircraft (F-16A/B and Mirage III/V) are nuclear capable with a range of 1,600–2,100km (1,000–1,305 miles) but the majority are carried on land-based ballistic missiles with a range of up to 1,500km (930 miles). The latest Shaheen-3 will have a range of 2,750km (1,710 miles) from its introduction in 2018. Pakistan has 12 nuclear-tipped cruise missiles with a range of 350km (217 miles). Missiles and cruise weapons carry nuclear bombs with a yield of 5–12KT.

Pakistan currently employs six ballistic missiles and is likely to be adapting the Chinese JF-17 to have nuclear capability. Potentially the most threatening are Pakistan's road-mobile missiles capable of rapid launch after redeployment. There are also clear indications that Pakistan could soon join the United States, Russia, China and India in boasting a nuclear triad of land, sea and air systems for nuclear deterrence; intelligence information from a variety of sources indicates the imminent availability of a submarine-launched ballistic missile programme for Pakistan.

Israel

Very soon after it was founded as an independent state, Israel began preparations for acquiring nuclear weapons. Facing political and military hostility from surrounding Arab states, it sought technical help from several European countries and engaged the cooperation of France in developing a robust industrial base for the economy. This association served it well when France itself began to acquire nuclear capabilities.

In 1949 a special military unit began surveying the Negev desert for possible uranium deposits and some were located in phosphate deposits. The Israel Atomic Energy Commission was formed in 1952, with its chairman David Bergmann pursuing his advocacy for an Israeli bomb, claiming that with this weapon the people would 'never again be led as lambs to the slaughter'. Bergmann also headed a group known as Machon 4, which perfected the means of extracting uranium from the desert and advanced the process of producing heavy water which gave Israel a new and indigenous capability for weapons production.

Prime Minister David Ben-Gurion was passionate about Israel acquiring nuclear weapons and the 1948 Arab–Israeli war only intensified the determination to acquire them. Internal support was strong and sound, with universal agreement within all political parties. Uranium extraction through phosphate excavated from the Negev was a unique process and became a bridge to work already under way in France, which resulted in cooperation between these two countries.

Responding to Eisenhower's Atoms for Peace speech, Israel signed up to the deal in 1955 that allowed the Americans to construct a small reactor in Nachal Soreq, which veiled development of a larger reactor at Dimona provided largely with the help of the French. The French had hoped to gain access to Jewish physicists who had been so prominent in launching America on the road toward the world's first atomic bomb. Israel willingly cooperated with the French, making the results of their knowledge gained through the Manhattan Project known to both countries.

A formal agreement in two parts, secret to

this day, was signed by France and Israel on 3 October 1957 which allowed for the provision of a publicly declared 24MW EL-102 reactor. The one actually built was many times that size and provided for production of up to 22kg (48.5lb) of plutonium each year. Aware of its true purpose, the public statements of Ben-Gurion that it was for powering a desalination plant to transform the desert into a flowering garden incensed scientists who prophesied that it would 'unite the world against us'.

By now there was a need to scour the globe for technical advisers, scientists and nuclear engineers, the search being placed in the hands of a new but secret government agency under Shimon Peres, but after de Gaulle

ABOVE Israel's first Prime Minister, David Ben-Gurion (1886–1973) addresses the Knesset in 1957. A powerful and enthusiastic advocate of acquiring nuclear weapons, he was driven by a desire to see Israel secure. *(Israeli Embassy)*

Declassified KH-4 CORONA November 11 1968

RIGHT Observed through the camera of a KH-4 spy satellite, Israel's Dimona nuclear reactor is key to its development of indigenous nuclear capabilities, which it veils in the most heavily guarded secrecy. *(CIA)*

became President of France relations between the two countries soured and he refused further assistance in supplying uranium until Israel opened its programme to international inspection. Not before 1966 did cooperation end, halting many years of clandestine and illegal supply of materials to Israel, including highly enriched lithium-6 for fission boost and as a fuel for hydrogen bombs.

The full-scale production of nuclear weapons has its origin in the days after the June 1967 Six-Day War, although by this date Israel had its first operational bomb. To support a major production programme the Israeli secret intelligence organisation Mossad obtained 200 tonnes of uranium ore from a Belgian mining company, supposedly to supply an Italian chemical company in Milan. But Israel was not averse to other procurement activity, including the theft of highly enriched uranium from a US Navy nuclear plant, according to the CIA.

The full story of Israel's secret nuclear weapons programme is fraught with uncertainties, rumour, obfuscation and disinformation. Suffice it to say here that the country has chosen a deliberate route of non-disclosure through its association with the

BELOW Rarely does Israel exhibit hardware even remotely associated with its nuclear arsenal. Here, the third stage of the Shavit satellite launcher is displayed bearing the sign of its manufacturer, Israel Aircraft Industries. The launcher relies for a lot of its technology on the two-stage Jericho II nuclear missile. *(IAI)*

South African bomb, which may have resulted in a single test on the South Atlantic on 22 September 1979.

Speculation within an environment of absolute secrecy has fuelled debate as to the precise number of nuclear weapons stockpiled by Israel and most intelligence assessments believe that there are approximately 80 devices capable of being carried on aircraft and missiles. Most prominent in terms of threat threshold are the Jericho series of short-range and long-range ballistic missiles. Jericho I appeared in 1971 and was specifically designed to carry a nuclear warhead. It had a range of 500km (310 miles) and was accurate to within 1km (3,300ft) with a warhead mass of 400kg (880lb). The missile was developed with assistance from France but with refinements from Israeli technology for the 100 or so produced.

Jericho I was taken out of service in 1990. Its replacement, Jericho II, is a solid-propellant, two-stage, long-range missile with a range of 1,400km (870 miles) and, potentially, a 1MT warhead. Ironically, this missile design originated during a cooperative venture with Iran before the overthrow of the Shah turned that country into a sworn enemy of Israel. Jericho II became operational around 1989–90 and forms the core of Israel's strategic nuclear deterrent. More recently Israel has completed development of its Jericho III, a three-stage solid propellant design that could be capable of carrying a single megaton warhead or several MIRV re-entry vehicles across a range of up to 6,500km (4,083 miles).

The Jericho III can be installed in very deep silos capable of withstanding a direct attack from a fissile-yield bomb and still retain its integrity for use, but there are certain indications that none of the Jericho series are on constant alert like those of India and Pakistan, which are held in a stockpile ready for use when the international situation warrants.

Although Israel is believed to have less than 100 nuclear weapons in total, with an ability to deploy these on aircraft, surface-to-surface missiles and possibly cruise missiles aboard ships, its nuclear weapons are arguably more survivable in the event of conflict, an imperative driven by the urgency of rapid response to some border countries with declared ambitions to completely destroy the nation as an independent state.

North Korea

Shortly after the Korean War of 1950–53, which ended in an armistice and which has never matured to a peace treaty, the North sought stronger ties with the Soviet Union, and in 1956 Russia started training scientists and engineers on how to start a programme leading to nuclear weapons. Provoked by the deployment by the US of its 280mm atomic cannon and by nuclear-tipped Honest John missiles to South Korea, the North signed a formal deal on cooperation with Russia in 1959. Three years later the Yongbyon Nuclear Scientific Research Center opened and by 1965 had reached a rating of 2MW, doubling that by 1974.

Beginning in 1980, several years after North Korea had started uranium mining operations in Suchon and Pyongsan, North Korea broke ground for a new factory at Yongbyon to refine yellow cake, a uranium concentrate powder that can be used for reactor fuel. By 1984 they were separating plutonium from spent nuclear fuel and had completed a 5MWe gas-cooled, graphite-moderated nuclear reactor and had started work on another 50MWe facility.

With the collapse of the Soviet Union, North Korea lost its prime sponsor and a programme of clandestine acquisition, of scientists, engineers, materials and knowledge, began to circulate around those countries seeking a nuclear weapons capability. In 1992 the International Atomic Energy Authority made an inspection of North Korea's facilities and judged that a clandestine weapons programme was under way and that far from its nuclear research being for electrical power, it was producing plutonium for warheads.

The first test occurred on 9 October 2006 when North Korea detonated a plutonium device in an underground test at P'unggye-yok delivering a yield of 0.2–1KT, attracting worldwide condemnation and US Security Council Resolution 1874 with sanctions which North Korea claimed would be regarded by it as a declaration of war. The country formally declared that it had nuclear weapons and on 8

October 2009 forbade international inspectors from entering the country.

A second underground test took place on 25 May 2009 with an estimated yield of 2–7KT followed by a third test on 12 February 2013 which seismic recordings indicated had a yield of about 7.75KT. Following further seismic recordings on 6 January 2016, North Korea claimed it had tested a thermonuclear fusion device but few analysts believe this, the test probably involving a boosted fission shot, albeit well on the way toward a true hydrogen bomb. A fourth test on 6 January 2016 was similarly claimed to have been a fusion test but this had an estimated yield of 7.1–15.5KT.

A fifth detonation on 9 September 2016 was estimated by Chinese experts to have delivered a yield of 11.9–23.7KT, bracketing the estimates from other countries of 20–30KT. Coming within a nest of missile flight tests, North Korea claimed that it now had the ability to produce a warhead and to marry that to a production missile but this claim was considered by many to be an exaggeration, despite the undoubted progress the country was making toward an operational system.

By this time China was one of North Korea's prime allies and suppliers, having replaced Russia as its main provider. But the highly secret programme, attached to belligerent and provocative statements regarding the willingness of North Korea to use nuclear weapons, meant China began a series of sanctions, banning all coal imports from that country, effective from 18 February 2017. China does not want conflict

ABOVE Honest John artillery rocket with a range of 24.8km (15.4 miles) on parade with South Korean forces in 1973, the presence of which was said by North Korea to be an unacceptable provocation. First deployed by the Americans in early 1953, Honest John was the US's first nuclear-capable surface-to-surface missile. *(US Army)*

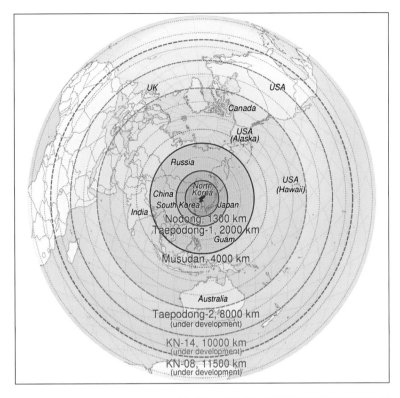

in the region; for purely pragmatic reasons it cannot afford to completely alienate the West and particularly the United States, which it sees, nevertheless, as resolutely committed to the defence of South Korea.

Strategic nuclear weapons are the very embodiment of North Korea's defence and power status and it is only a matter of time before it has a credible deterrent force. At present it probably has about six warheads, but the transition to a credible delivery system is still some way off. Over the last few years North Korea has engaged in a robust and escalating series of tests with long-range missiles, achieving varying degrees of success. The much-lauded KN-08 (Hwasong-13), which has a maximum range of 5,500km (3,420 miles), is still far from operational.

If North Korea can master the miniaturisation of warhead design, achieve a reliable means of arming and detonating a nuclear device, and can successfully integrate those elements into a survivable re-entry warhead, it will begin to achieve the weapon system it so desperately seeks. But the omens are not good for near-term success. Between 6 January and 20 October 2016 North Korea launched 24 missiles on test of which 11 failed. The longest range in that year was achieved by the No Dong weapon fired across a distance of 800km (500 miles).

CENTRE South Africa's RSA-3 missile which was developed with Israel, ostensibly as a satellite launcher but effectively to operate as a nuclear-tipped ballistic missile. When South Africa renounced nuclear weapons it attracted high levels of interest for its commercial value as a civilian satellite launcher. *(Via David Baker)*

LEFT A uniquely close working relationship existed between Israel and South Africa during development of the RSA series of missiles and satellite launchers, the initial RSA-1 being a copy of the Jericho II second stage which was adapted into a single-stage missile. *(Via David Baker)*

Aspirant powers

South Africa was quick to take up the American Atoms for Peace programme by signing a deal in 1957 allowing it to acquire a single nuclear research reactor with the aim of using this to provide electrical power to the grid. Over the next 20 years it acquired the means to produce its own fuel and to begin the design of a nuclear weapon. In 1977 South Africa chose a region in the Kalahari Desert at the Varstrap weapons range for test detonations, and preparations were made for a cold test that August. It did not happen.

Soviet intelligence alerted the US, which obtained data from a Lockheed SR-71 reconnaissance aircraft to verify that this was so. France threatened to cancel all its agreements relating to civilian nuclear power facilities, general pressure was applied to South Africa and the test was cancelled, the holes being quickly filled in. The plan had been to use aircraft such as the English Electric Canberra B12 and the Hawker Siddeley Buccaneer to deploy the bomb along with missiles such as the RSA-3 and RSA-4. But ambition outstripped capability and South Africa was a very long way off matching bomb designs to delivery platforms.

CENTRE Preserved in a museum, a Blackburn Buccaneer of the type assigned to carry tactical nuclear weapons when South Africa attempted to become a member of the Nuclear Club.
(Via David Baker)

RIGHT Iraq's Scud missiles, that most prolific of Soviet-era weapons exports from the USSR, threatened countries across the Middle East, and during the Cold War the Soviet Union did not desist from threatening to equip those operated by its client states with nuclear warheads. (DoD)

But work continued in high secrecy, with the first highly enriched uranium being produced in 1978. Some degree of cooperation with Israel allowed an exchange of nuclear materials, South Africa trading 50 tonnes of uranium for 30gm of tritium. But the ownership of a nuclear detonation observed from space on 22 September 1979 is still unresolved, if it was nuclear at all, which has led to some speculation.

The 'observation' was made by a Vela satellite, which recorded a double flash of precisely the type in intensity and duration that would be produced by a nuclear detonation of 1–2KT. Several government agencies and private analysts have concluded that it was an instrument anomaly on the satellite but

the Vela series observed 41 such events by other countries, all of which turned out to be accurate. Moreover, subsequently a Soviet intelligence source told this writer that the event was a detonation code named Operation Phoenix conducted by Israel and South Africa.

Whatever the truth, South Africa continued to work toward a nuclear capability and had produced the first deliverable bomb by 1982, followed five years later by the start of a production run during which seven complete bombs were assembled. A test detonation was prepared in 1988, but this was cancelled, further production was stopped and the existing bombs were dismantled. In 1991 the country signed up to the Nuclear Non-Proliferation Treaty (NPT).

Other countries too have aspired to possessing nuclear weapons, including Iraq. During the early 1970s, Saddam Hussein ordered development of a nuclear weapons programme and manipulated the International Atomic Energy Commission into approving the importation of nuclear power technology for peaceful purposes. This even extended to acceptance of an unsolicited offer from the naïve Commission that Iraq should start a plasma physics programme for the peaceful use of fusion research – the perfect cover. Progress was halted in 1981 when Israel carried out a pre-emptive bombing raid on a huge and expanding range of facilities at the Al-Tuwaitha covert enrichment site.

In 1975 Iraq secured an agreement from Russia for the provision of an advanced reactor and France sold it 72kg (159lb) of 93% enriched uranium. When Saddam Hussein assumed total power in 1979 the country was in an almost total state of war, restrictions being placed on the movement of people including its nuclear scientists and engineers who by this time numbered several thousand.

A wide range of European countries and companies were recruited to help Iraq through the trading of myriad items, no one of which was clearly related to the building of an atomic bomb. In the end it was easier for the regime to turn to gas and biological agents and the nuclear programme evaporated with the war with Iran (1980–88), the Gulf War (1991) and in the invasion of Iraq by coalition forces (2003).

Ironically, the Middle Eastern country most watched in terms of nuclear weapons potential is the one that has made the loudest public statements for a nuclear-free zone in the region – Iran. But despite that there is unequivocal evidence that Iran aspires to become a nuclear-weapons power and that it is edging closer to acquiring bombs. Sympathisers would point to the Israeli monopoly on nuclear weapons in the region, on the fragility of peace with its enemy Iraq and on the protective conventional umbrella of the military might of the United States; cynics would point to the regional power struggle where Iran tries to rid Israel of its nuclear weapons through an international agreement using foreign countries as proxy negotiators to lower Israel's threat potential.

For its part, Iran points to its signature to the NPT and in 2015 entered into an international agreement whereby 20% of its plutonium would be removed from the stockpile, the stock of low-enriched uranium would be reduced from 7,154kg (15,775lb) to 300kg (661lb) and the complete removal of its medium-enriched uranium, then standing at 196kg (432lb). A very tight and comprehensive series of constraints and restrictions were agreed to and Iran has been brought back from the brink. However long that remains the case largely depends on the fragile peace that exists between that country and its enemies.

CENTRE Iran's Sejil-2 missile (left) and Qiam, the latter having entered service in 2011 with a medium-range capability and a guidance package that can, it is claimed, manoeuvre the stage to avoid anti-missile systems. The Sejil-2 MRBM has a range of 2,000km (1,240 miles). *(Via David Baker)*

RIGHT One of several Iranian missile firings as that country works toward a medium- and long-range missile capability. *(Mahmood Hosseini)*

Chapter Eight

The big stick

The technology to deliver nuclear weapons to their intended targets was almost as challenging as the development of the weapons themselves, with several breakthroughs needed to provide nuclear powers with reliable and effective delivery platforms.

OPPOSITE A Northrop Grumman B-2 Spirit flies across St Louis, Missouri, a strident and potent reminder of US nuclear strategic power. *(USAF)*

The 26th President of the United States, Theodore Roosevelt, was known for many things, not least his love of country, freedom of will and the natural world, but he is perhaps more frequently quoted for his template for all foreign policy initiatives: 'Speak softly, and carry a big stick!' That edict could be the motto of the nuclear age, an approach to challenges and threats in which, as Roosevelt said, preparedness was based on the 'exercise of intelligent forethought and of decisive action sufficiently far in advance of any likely crisis'. It was so in the decision to create a nuclear 'big stick', to serve as a reminder of potential retribution by punishing an aggressor with unacceptable and disproportionate use of force.

Initially atomic bombs were air-dropped devices utilising conventional body shapes developed from the various designs of conventional bombs employed during the Second World War. To some extent the physical shape and proportionality of the bomb was dictated by the configuration of the early bombs and by the element selected for the fission process.

The now familiar shape of the missile nose cone was defined by two discoveries: that a blunt body fabricated from known materials would provide the optimum shape for surviving the heat of re-entry; and that it would provide maximum deceleration by optimising lift over drag. Perhaps surprisingly, it was also due, in no small measure, to the miniaturisation of nuclear weapons.

The story of why that was so is an integral part of understanding why the blunt-body shape is optimum for getting anything back from space intact. It began with the development of atomic weapons in the 1940s, research which itself led to developments with calculating machines and computers that provided opportunities for physicists and nuclear weapons engineers to more fully understand what happens in the chaotic behaviour of natural forces, the key to understanding how nuclear weapons work.

Early atomic bombs designed to liberate energy from fission were very heavy and quite large. Very few aircraft had the capacity to carry them until the advent of intercontinental bombers such as the Convair B-36. Operational with the US Air Force Strategic Air Command between 1948 and 1958, more than 400 of these ten-engine giant bombers were built, the only operational aircraft capable of carrying every US nuclear weapon in the inventory.

The size and weight of atomic weapons, and the more powerful thermonuclear weapons based on fusion which were developed in the early 1950s, was a restraint on the early development of very long-range missiles. The rocket engines of the 1940s and '50s were incapable of providing the power required to propel a large rocket to sufficient velocity and this is why intercontinental ballistic missiles (ICBMs) were not considered viable until a major breakthrough occurred in nuclear weapons design.

The early atomic weapons were either spherical or torpedo-shaped, but in 1952 – the

year in which the first thermonuclear (hydrogen) device was tested – a series of theoretical and practical developments allowed major improvements in the physical engineering of the primary stage of a nuclear weapon, reducing the diameter by a factor of three and the weight by a factor of 30. By 1954 these developments had been fed across to nuclear weapon applications as diverse as artillery shells, tank rounds, small bombs dropped by fighter-bombers, sea mines and air-air-missiles fired by interceptors against massed bomber formations.

By 1956 the theoretical possibilities were seemingly limitless. The Atomic Energy Commission indicated that a 60MT bomb was possible and the US Air Staff were told that a yield of 1,000MT might be plausible, a device with 50,000 times the power of the bomb used at Hiroshima. But instead of thermonuclear weapons getting bigger and more powerful, the emphasis switched to making them smaller, less powerful but much more accurate. The decision in 1954 to develop ICBMs on a crash basis was facilitated by the availability of these warheads, but the need for survivable re-entry vehicles was pivotal to their value as a deterrent. The peak period of building a tactical nuclear air force ran between May 1951 and July 1953, the latter immediately preceding the decision to produce Atlas, and eventually Titan, ICBMs.

While the Air Force retained its own research and engineering development structure the basic theoretical work that would lead to a satisfactory means of bringing a warhead back through the atmosphere intact fell to a civilian organisation that had been established by the government in March 1915. The National Advisory Committee for Aeronautics (NACA) was set up in response to expanding aeronautical capabilities in Europe. With just 12 personnel, it was required to carry out essential advisory functions but would quickly grow into a major research and engineering development body.

When America entered the First World War in April 1917 the need for a more formal technical development organisation became apparent and the Langley Memorial Aeronautical Laboratory was built at Langley Field, Virginia, from where vital research was conducted into a wide range of technical challenges to aircraft design, including aerodynamics, wing section profiles, propeller design and materials technology. Responding to the surge in aeronautical design during the 1930s, and the advent of military threats in Europe, in 1939 NACA set up its second research facility.

Named after NACA's venerated chairman from 1929 to 1939, Dr Joseph S. Ames, the Ames Aeronautical Laboratory was located at Moffett Field, California, close by Sunnyvale, on a boggy piece of land 61km (38 miles) from San Francisco at the foot of the Bay. As the former home of the US Navy's unsuccessful hopes for a family of rigid airships, it was redundant to purpose and made an ideal place to set up an aeronautical research facility.

But in its role as a research organ of the government, NACA applied its facilities,

BELOW The various components of the Minuteman ICBM and the various contractors which contributed to the fabrication and assembly of the US land-based missile leg. *(Via David Baker)*

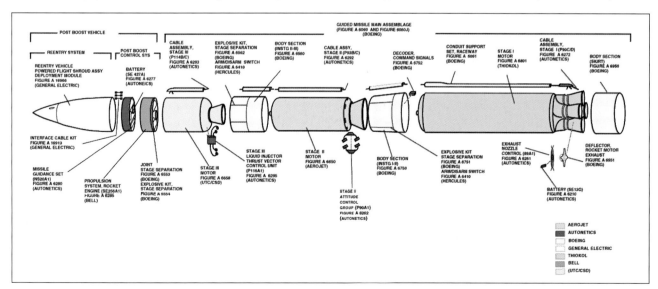

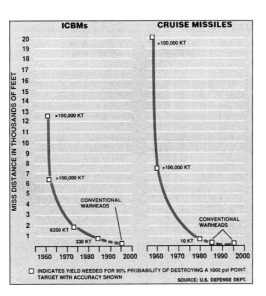

laboratories and test sites to a wide variety of scientific and engineering problems outside the general area of flight and flying. Among these were finding solutions to problems involving very high-speed flight in and out of the atmosphere. Just as NACA was thinking about hypersonic

flight research, bridging the zone between aerodynamics and thermodynamics, some researchers at the Ames Aeronautical Laboratory were studying the problems encountered by high-speed aircraft (projectiles) such as the forthcoming X-15 hypersonic research aircraft.

Using specially adapted wind tunnels at Ames, NACA engineers sought solutions to aerodynamic heating on a body travelling through the atmosphere at great speed. Undertaken as part of essential research into what would popularly, and incorrectly, be termed the 'heat barrier', this work attracted the interest of engineers seeking solutions to the problem of getting re-entry warheads back through the atmosphere. It was exciting work, because a solution would be pivotal to high-speed flight.

Rockets with even modest range had an apogee on their trajectory outside the atmosphere but at velocities where kinetic energy on re-entry did not produce significantly high temperatures. The V-2 is one example: by 1943 V-2 rockets were arching over to a peak altitude at the fringe of space. But rockets designed to throw a warhead several thousand kilometres would fully exit the atmosphere and reach altitudes as high as 1,450km (900 miles), achieving velocities as high as 19,300kph (12,000mph). The stagnation temperature in the shock wave of such a body would reach 6,650°C (12,000°F), about ten times the temperature that would be experienced by hypersonic aircraft such as the X-15, then in development.

Early rockets like the V-2 had warheads integral with the forward section of the projectile, but instability caused by the significant change in its centre of mass as its propellant was depleted caused tumbling. This frequently resulted in the destruction of the entire vehicle as it re-entered the denser layers of the atmosphere, warhead included. Warheads which could be made to separate from the main body of the rocket after burnout were the solution, leaving the spent tankage and rocket motor to be destroyed during re-entry. But this still left the problem of how to make a warhead capable of surviving the extremes of heat through atmospheric friction.

The solution evolved quite rapidly at NACA's Ames Aeronautical Laboratory, where Harvey

BELOW The Minuteman III missile is America's land-based ICBM, comprising the first stage (A), second stage (B), third stage (C), post-boost vehicle, or PBV (D), which previously carried three MIRV warheads, and the payload shroud (E). A typical flight sequence begins with launch out of its silo (1), separation of the first stage and ignition of the second stage 60 seconds after lift-off with ejection of the payload shroud (2), ignition of the third stage two minutes after launch (3), with third stage burnout and release of the PBV 60 seconds later (4). The PBV manoeuvres (5) and prepares to re-enter, deploying separate warheads and chaff (6) before the warheads are armed (7) and speed to their separate targets (8). (David Baker)

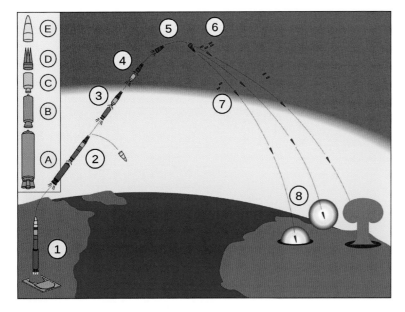

Julian Allen was challenging the idea that a pointed nose shape was the most streamlined and therefore the most effective way to penetrate the atmosphere, which it was if that object were travelling at supersonic, not hypersonic speeds of re-entry from space. Already, aerodynamic heating of the exterior surface of very high-speed aircraft was becoming a problem.

By the late 1940s Allen had a fairly well-placed knowledge of how objects of various shapes and configurations behaved from subsonic to high supersonic speeds and his interest was extended further when dedicated studies were made into the appropriate shape of a ballistic re-entry vehicle carrying a nuclear warhead. He was well aware that the kinetic energy lost by a warhead would be totally converted to heat and that this energy would come from two places: inside and outside the boundary layer. Some of the heat in the boundary layer would result from compression of the air but the higher proportion would result from viscous shear or skin friction.

Allen was motivated by a belief that the so-called 'heat barrier' of the 1950s was as mythical as the 'sound barrier' had been in the mid-1940s and that a solution could be

found to the problem that no known materials could withstand the temperatures incurred by adopting conventional design methods for aerodynamic streamlining. Allen based his solution on the fact that heat in the shock wave outside the boundary layer was largely isolated from the body itself and could not reach it through convection due to the boundary layer acting as an insulating blanket.

Allen calculated that what he referred to as 'pressure drag' could take away a lot of the thermal energy if a larger fraction of the heat were generated in the shock wave itself, reducing the heat level in the re-entry body. While the overall level of heat generation would depend on the degree of deceleration, the heat produced in the boundary layer must be as turbulent as possible and not laminar. By increasing pressure drag, a larger portion of the thermal energy would be generated in the shock wave than by viscous skin friction.

Because the bow wave in front of the decelerating object is more powerful than the oblique waves angled back from the pointed nose it generates much more heat but it does not come into contact with the body at any point. The oblique waves come into contact with the decelerating body where there is a

the balance between deceleration and lower thermal loading on the surface, also minimising hot spots. It was also beneficial in reducing skin friction by maintaining laminar flow over the heated area. But there was a lingering concern that the maximum heat rate would not be the same for a blunt body as for a pointed one.

Along with Alfred J. Eggers, Allen demonstrated through mathematics that the maximum rate at which the body generated heat was determined by the entry angle and the velocity, irrespective of shape, size, mass or drag coefficient. The same amount of heat would be generated by both a blunt and a pointed body but the amount of heat that actually entered that body would be much less with a blunt shape, base forward. They also discovered that the maximum deceleration would occur at 61% of the entry velocity and that this held true irrespective of flight-path angle.

By the end of 1951 the work was completed and the theory had been tested in a variety of ways but the report was highly classified, it being conveyed on 28 April 1953 first to the scientists and engineers at work on the ballistic missile programmes before being eventually appearing as NACA TN 4047 and then, in open literature, as TR 1381 on 1 January 1958. In this document Allen and Eggers summarised by asserting that 'A shape which should warrant attention for such missile applications is the sphere…'.

Its logic resulted in the development of General Electric's Mk I and Mk II warhead designs for the Atlas missile, and the shape was tested and applied to the Thor missile. Married to the significant reduction in weight from miniaturisation of thermonuclear warheads, allowing missiles of acceptable size to carry them across intercontinental distances, the ability to provide survivable warheads to carry the devices to their targets overcame the limitations that had dogged missile designers for a decade. Never given the credit they deserve, Allen and Eggers reshaped the mathematical basis of re-entry warheads.

As debate revolved around the use of re-entry vehicles, issues surrounding the type of materials preferred for the warheads flared up. Despite the Allen/Eggers blunt-body concept, temperatures would still be very high. At the start of re-entry a blunt body would

thin or non-existent boundary layer acting as an insulator. Although they do not carry as much thermal energy as the bow wave, the oblique waves convey their heat to the tip of the re-entry body with greater ease and at a speed that prevents it being carried away through conduction. This is why the tip rapidly heats up and melts the metal of which it is made.

It was this conclusion, that a blunt body experiences much less heat than a pointed one, that convinced Allen that a bowl-shaped re-entry, with low convex ratio, would optimise

create a shock wave with a temperature of 5,260°C (9,500°F), while the blunt face of what constituted a heat shield would be at a temperature of 1,650°C (3,000°F).

There were just two principles governing the category of materials that could safely protect warheads and incoming re-entry vehicles: heat sink or ablation. Existing materials such as aluminium and titanium were universally applied to the design of hot structures experienced by X-series NACA research aircraft, but new materials – such as Monel K and Inconel-X – were already being investigated as thermal protection for hot structures using the heat sink method. Ablative materials would char away and take with them the heat conducted across the boundary layer.

The Army was keen to test ablative materials and on 8 August 1957 a Thor rocket propelled a warhead protected by an ablative material to a height of 965km (600 miles) and a range of 1,930km (1,200 miles). Whether ablative materials would be effective for the much greater energies felt by an incoming warhead from an intercontinental trajectory was something the Air Force questioned. NACA was asked to test various ablative materials, and Teflon, nylon and fibreglass composites were tested as well as heat-sink materials such as copper, beryllium and tungsten. But the Air Force conducted its own tests using the X-17, a specially built research rocket from Lockheed Aircraft Corporation, using warhead materials from General Electric for heat-sink concepts and from Avco Corporation for ablatives.

Throughout 1955–56 NACA-Langley exposed a range of materials to temperatures of up to 2,260°C (4,100°F) in the plume of an acid-ammonia rocket jet at a gas stream velocity of 2,133m/sec (7,000ft/sec). High-temperature electric arc jets at Langley and at the Lewis Flight Propulsion Laboratory provided energy for compressed air and for increasing the pressure and temperature of the air, producing a steam gas temperature of 6,650°C (12,000°F). The results were given to the Army, vindicating its belief in ablative materials, however crude they appeared in principle.

But there are other factors that determine the accuracy of the warhead as it re-enters the atmosphere at hypersonic speeds. The aerodynamic qualities of the re-entry body will determine how it behaves in the atmosphere. Warhead designers, initially at least, were seeking a non-lifting re-entry body design which would not change course due to its shape.

The most important aspects are the atmospheric pressure, the Mach number and what is known as the Reynolds number. Smooth, or laminar, flow and the onset of turbulence are factors that can disturb the trajectory of an incoming warhead, and in the very early days of missile development accuracy was an issue for the propulsion and guidance systems of a rocket. The warhead was designed to follow a predictable path to the target, accuracy being governed by the velocity, pointing angle and altitude at cut-off.

Defined by Osborne Reynolds, Professor of Engineering at Manchester University in the 1880s, the Reynolds number has no dimension as such but is the ratio of inertial forces to viscous forces and is a fundamental parameter in fluid mechanics which governs the way flow patterns in a fluid depart from laminar flow to turbulent conditions. While the principle is simple in concept, the mathematics were one of the earliest derivations necessary for successful powered flight. It followed the design engineer through the history of flight as an essential parameter for determining the motion of a body in a fluid – either water (solid) or air (gaseous).

This parameter is an essential tool in creating

BELOW A test of the US ICBM system sends dummy warheads spearing into their simulated targets in the Western Pacific Ocean, effectively representing in this one image a collective destructive power greater than all the explosives detonated in war since the first, several centuries ago. *(USAF)*

ABOVE A distribution map of selected ICBM sites at three Missile Wing facilities attached to Malmstrom, Minot and Francis E. Warren Air Force bases, their distribution intended to prevent a single warhead from threatening more than one silo. *(USAF)*

BELOW The Trident SLBM is ejected by gas from the missile tube on a submerged submarine, after which it ignites its rocket motor to begin its trajectory to the target (images from bottom to top left). *(USN)*

TRIDENT I FLIGHT TEST

effective flow prediction in any solid body moving through a fluid and Reynolds discovered the underlying and unifying principle between the two. It is no overstatement to say that it is the most important quantity in the whole of aerodynamics and dictates the design of any physical structure designed to move through air or water. It is a non-dimensional quantity and as such is a pure number. This writer is at a loss to explain why it appears so infrequently in the literature on fast aircraft design, but for warhead design engineers it was a crucial value.

Many experiments were conducted, some by NACA, to determine the different Reynolds number for specific warhead designs. The unpredictable onset of turbulence could destabilise the post-boost trajectory and undo all the work done to position the rocket to deliver an accurate flight path to the warhead. Perturbations in the trajectory were not so important at the beginning, but as accuracy increased and theoretical miss distances tumbled these factors became very important and would even be utilised to create adjustable re-entry paths to null the dispersions of thrust termination at the end of the boost phase.

From the beginning of ballistic missile design, scientists were concerned about the effect of dust in the atmosphere and of possible interference from meteor showers, the presence of which is dictated by the passage of the Earth in its orbit through the fixed elliptical paths of cometary fragments. Assuming serious study in the early 1950s, long before the advent of direct measurement from satellites and space probes, relatively primitive techniques were used to discover the magnitude of the threat. Direct measurement had been possible only through the scientific research supported by ballistic, suborbital sounding rockets.

Studies from the ground of zodiacal light – a diffuse glow caused by sunlight scattered through clouds of space dust – indicated that there would be little or no induced deflection in the trajectory and that the only concern was of very high-speed impact with minute fragments possibly scarring the re-entry body encountered during high-speed descent through the atmosphere. But in the mid-1950s the only tools available for determining a precise value were the measurement of pits on recovered rockets,

the collection of micrometeorites, the study of deep-sea nickel deposits and the computational determination of optical density compared to the average diameter of dust particles.

In several ways the effects of re-entry and its associated perturbations from all these factors are influenced to a lesser or greater degree by the exterior ballistics of a missile and its warhead. After shutdown of the propulsion system the body is in free-fall to the target, the only force acting on it being gravity. The length and shape of the free-flight trajectory are determined by the speed of the missile at cut-off, the angle between the local vertical and the direction of velocity (the vector), the altitude at cut-off and the gravitational acceleration along its path from the time of cut-off.

Two extreme trajectories are possible, one that is steep with a high apogee and another that is shallow with a low apogee, although there are a range of options between the two. For the steeper trajectory the re-entry speed is higher, presenting a more formidable heating problem with high heat rate. For the flatter trajectory the re-entry path is longer with less of a peak thermal load. The amount of heat absorbed will be the same in both cases since the same amount of energy has to be dissipated. The extremes of these

ABOVE In October 1969, President Nixon resorted to the 'madman theory' in which he wanted the Soviet Union to believe that he would stop at nothing to end the aggression from North Vietnam. He slipped word to the Russians that he was at his wits' end and would use nuclear weapons to wipe out North Vietnam. Henry Kissinger authorised Operation Giant Lance and 18 B-52s from the 92nd Strategic Aerospace Wing, each carrying nuclear weapons, were sent racing toward the Soviet border openly observed by the Russians, only turning back within seconds of crossing the border. The mission was flown on 27 October and called off three days later. *(USAF)*

LEFT B-52s were on high states of alert throughout the Cold War, and during selected periods in the 1960s were frequently flown with nuclear weapons ready for war. On 24 January 1961, a B-52G broke up in flight and dropped its two 1.9MT Mk 39 nuclear bombs on Goldsboro, North Carolina. After deploying its parachutes five of the six arming stages were completed but the sixth held. *(USAF)*

RIGHT In this sequence of drawings, the effects of a nuclear attack are displayed for two different types of weapon, a fission bomb with a yield of 20KT and a thermonuclear device with a 1MT yield. The radial distance of the effect described is displayed at the bottom according to the type of bomb. Here, the gaseous fireball forms and emits thermal radiation followed by gamma rays and neutrons and an expanding blast wave. *(Samuel Glasstone)*

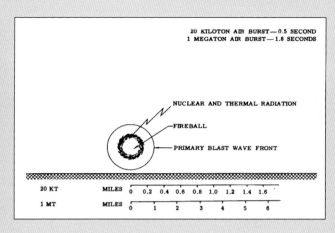

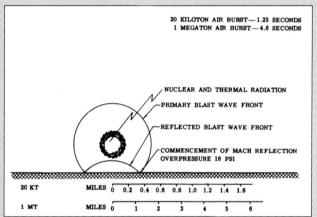

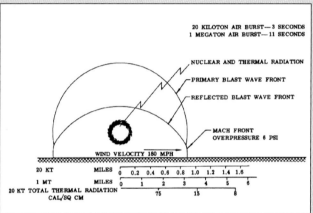

ABOVE The primary air blast strikes the ground and a secondary, reflective wave follows. At a prescribed distance depending on the height of the burst and the energy of the yield, they fuse to form a single reinforced stem travelling at supersonic speed. *(Samuel Glasstone)*

BELOW The effects ten seconds after detonation, where the fireball is no longer luminous but continues to rise upward, drawing air in and producing strong afterwinds, raising dust and dirt into the air to form into the stem of the mushroom cloud. *(Samuel Glasstone)*

ABOVE As time progresses the stem moves outward and increases in height with an overpressure at the Mach front of 41.37kPa (6lb/in²), with the wind behind moving at 290kph (180mph). The thermal radiation from a 20KT burst lasts only three seconds, while that for a 1MT burst lasts for 11 seconds. *(Samuel Glasstone)*

BELOW As the thermal residue rises it cools, and vaporised fission products and residue from the bomb condense into a cloud of highly radioactive material. The afterwinds reach a maximum velocity of 320kph (200mph). *(Samuel Glasstone)*

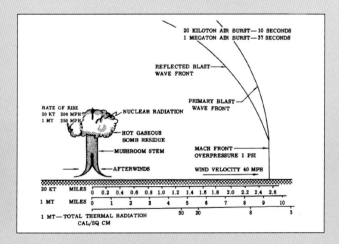

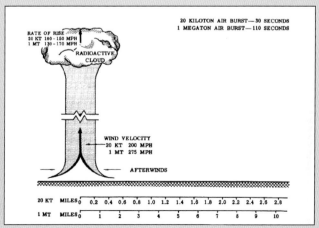

two trajectories require more demanding navigation requirements.

Another factor complicating the trajectory is the azimuth of launch as measured from the equator, which has a circumference of 40,075km (24,901 miles). The Earth is not a static body and revolves with an equatorial speed of about 1,670kph (1,037mph), decreasing with latitude due to the approximately spherical shape of this modestly oblate body. In much the same way as a marksman shooting at a moving target has to calculate the degree of deflection to compensate for the time taken for the discharged bullet to reach its target, so too must the rocket be fired so that it will come down on its target after accounting for the rotation of the Earth in the time between launch and impact.

When striving to achieve orbit, ideally the rocket will fly to space along the line of the equator perpendicular to the axis of rotation. If launched toward the east at 0° inclination, the rocket will gain an advantage from the spin of the Earth – anticlockwise when viewed downwards from directly above the North Pole. The fixed velocity required to achieve orbit will be reduced by the speed of the rotating Earth already moving the rocket eastward before it is launched. Launching west along the line of the equator means that the rocket must make up the speed of the Earth going in the opposite direction before adding to its required total energy the velocity required for orbital flight.

For suborbital, ballistic flight, however, none of that matters, but the rotation of the target fixed to the surface of the Earth will appear faster up to a flight path at 90° to the equator and then decline again. This is best illustrated by considering a spacecraft launched into polar orbit at 90° to the equator. In taking 90 minutes to make one complete revolution of the Earth the fixed point on the surface from which it departed will have rotated east approximately 2,500km (1,555 miles). This is why the Space Shuttle had a large wing area, to be able to 'turn' the incoming trajectory to fly eastwards if it had to come down after one orbit of the Earth.

Ballistic missiles too have the same problem when fired to a target that is moving. The second the rocket leaves the pad the projectile is fixed in its trajectory relative to the centre of the Earth's mass and must fly to a target that is constantly moving in an eastward direction. The complexity of the calculation can be appreciated when considering a variety of different targets from different launch sites, each with its unique rotation rate relative to the motion of the Earth on its axis.

Moreover, the Earth is not a homogenous agglomeration of rocky materials and massive oceans, and the force of gravity is different at various locations. Accordingly, the trajectory calculated by the flight path engineer must take account of these anomalies so that the forces acting on the rocket and on the unpowered warhead will not shift the projectile off course. An accurate geodetic mapping of the Earth's gravitational field is vital, and more so with enhanced accuracy, which is why it was one of the first determinations sought from special geodetic satellites, each measuring the discontinuity of attraction with the centre of the Earth so that the variations in mass concentration could be factored into the guidance equations.

BELOW The stealthy, nuclear-armed B-2 was developed to enhance its survivability inside the Soviet Union, hunting down mobile ICBMs that it was believed the Russians were developing. That same capability is now adapted to seek out military targets inside heavily defended airspace, where its low radar cross-section enhances survivability, increasing its ability to escape detection. *(USAF)*

Bans and denuclearisation

Arms agreements and limitations have brought adversaries together in an international effort to eliminate the harmful products of the nuclear age and to place restrictions on proliferation while constraining the use of harmful and radioactive materials.

The United States assembled more than 70,000 nuclear weapons during the Cold War and at the end it had 23,000 weapons of 26 different devices. Arms Agreements such as the Strategic Arms Reduction Talks II signed in 1993 had anticipated a reduction to between 3,000 and 3,500 devices but it never came into effect. Subsequent agreements committed to a lowering to 1,550 warheads each for Russia and the United States.

After the collapse of the Cold War, in 1993 an agreement was reached between the United States and Russia for conversion of weapons-grade highly-enriched uranium (HEU) to low-enriched uranium (LEU) for use in power stations. Russia agreed to convert 500 tonnes of HEU to LEU, enough for 20,000 warheads, which was bought by the United States for use in civil nuclear reactors. This was achieved through removal from the warhead, before being machined into shavings, oxidised and fluorinated. This produces a highly enriched uranium, hexafluoride, which is mixed with slightly enriched uranium in a gaseous stream, forming LEU for the reactors. It is transferred to shipping containers and sent to a collection point in St Petersburg. A supplementary agreement was signed in 1999 to cater for uranium feed from mines which was blended down and sold to the customers.

The US government declared 174 tonnes of HEU as surplus to its military needs and the international agency handling exchanges took possession of 134 tonnes of uranium hexafluoride containing about 75% U-235, and 50 tonnes of uranium oxide containing 40% U-235. In 2005 a further 200 tonnes was declared surplus to military requirements, of which 160 tonnes was to be retained for nuclear propulsion in submarines, postponing the need for a uranium high-enrichment facility for 50 years. An additional 20 tonnes was set aside for use in nuclear generators aboard space missions and for research reactors.

The general downsizing of the Russian and American nuclear arsenals resulted in about 150–200 tonnes of weapons-grade plutonium and in 1994 the USA declared that half its

BELOW President John F Kennedy signs the Nuclear Test Ban Treaty on 7 October 1963, ending all atmospheric testing by the superpowers. *(The White House)*

military plutonium stockpile was surplus to requirement. In June 2000 both the USA and Russia agreed to dispose of a minimum 34 tonnes each of weapons-grade plutonium. The US decided to self-fund that but the G7 agreed to step in and provide $2.5 billion to set up a Russian programme. It was agreed that some of this could be used in Russia's fast neutron reactors and the timeline was stretched to 2018. The adaptation of the initial agreement meant that Russia would fund the programme itself, to the greater benefit of verifiable monitoring through inspection. The 68 tonnes of plutonium thus removed from potential weapons application is worth about 12,000 tonnes of natural uranium.

By 1998 the UK had an estimated 7.6 tonnes of military plutonium, slightly less than half of which was of weapons-grade quality. About 4.4 tonnes of the total was declared surplus by the British government in addition to 100 tonnes of reactor-grade plutonium for uses which are yet to be determined.

Progress in eliminating weapons-grade materials for nuclear weapons has made tremendous progress. In June 2013 the National Nuclear Security Administration (NNSA) declared that it had monitored, and verified, the complete elimination of 475 tonnes of Russia's HEU under the Megatons

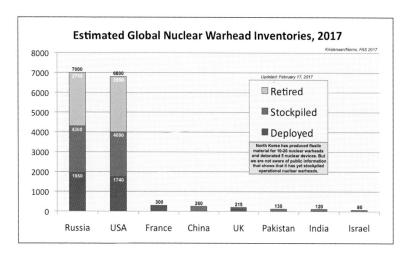

to Megawatts programme. The agreement to produce LEU from HEU was 95% complete, the original material being the equivalent of 19,000 nuclear weapons that have been permanently eliminated.

Under the umbrella of intensive work by both Russian and American monitoring teams, working with each other to dismantle not only the weapons but the means to provide them with warheads, the Russians have accepted visits on 359 occasions between 1995 and 2013, a transparency programme unique in its effectiveness and productivity. For their part, the Americans have opened up their own facilities to Russian inspection, to the total satisfaction of both sides.

ABOVE Current inventories with indicated numbers of estimated weapons showing the considerable downsizing of US and Russian forces.
(Via David Baker)

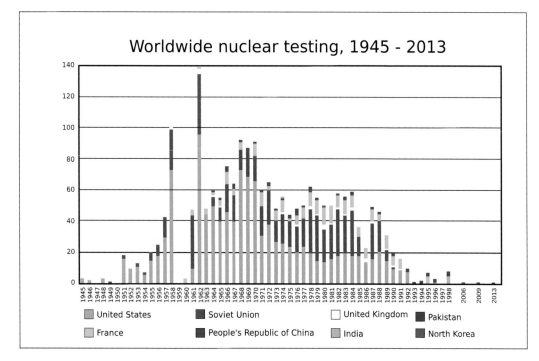

LEFT Worldwide nuclear weapons testing 1945 to 2013.
(ACDA)

Abbreviations and glossary

A-bomb – Atomic bomb.

ACDA – Arms Control & Disarmament Agency.

ADM – Atomic demolition mine.

AEC – Atomic Energy Commission.

Alpha-decay – Process in which a nucleus loses a binding pair of two protons and two neutrons, which lowers the mass number by four units but the atomic mass by only two (through the departing protons).

amu – Atomic mass units.

ARL – Airfield Radio Laboratory.

ASMP – Air-sol moyenne portée; medium-range air to surface missile..

ASMP-A – Air-sol moyenne portée-amélioré ('improved ASMP').

AWE – Atomic Weapons Establishment.

AWRE – Atomic Weapons Research Establishment.

BAOR – British Army of the Rhine.

BARC – Bhabha Atomic Research Centre.

BE – Binding energy.

Bepo – British Experimental Pile-0.

Beta-decay – Process that occurs when one proton in the nucleus becomes a proton by ejecting a single unit of negative charge or an electron.

BJP – Bharatiya Janata Party.

BMEWS – Ballistic Missile Early Warning System.

BNFL – British Nuclear Fuels Limited.

C – Coulombs.

CASC – China Aerospace Science and Technology Corporation.

CASD – Continuous at-sea deterrent.

CEA – Commissariat à l'Energie Atomique.

CENTO – Central Treaty Organisation.

CEP – Circular error probability, the radius within which 50% of missile rounds will fall.

C/gm – Coulombs per gram.

CIA – Central Intelligence Agency.

CIRUS – Canadian-Indian Reactor Uranium System.

cm – Centimetres.

Criticality – The level at which a reactor becomes capable of producing a predetermined level of power.

CTBT – Comprehensive Test Ban Treaty.

DIA – Defense Intelligence Agency.

dyne – Ten micronewtons.

Electroweak force – Integration of electromagnetic force and weak nuclear force.

EMIS – Electromagnetic isotope separation.

EMP – Electromagnetic pulse.

ENDC – Eighteen-Nation Disarmament Conference.

ERW – Enhanced Radiation Weapon.

Fat Man – Early implosion-type nuclear bomb.

ft – Feet.

ft/sec – Feet per second.

g/Mwd – Grams per megawatt-day-thermal.

Gadget – Colloquial name for the implosion-type plutonium bomb used for the Manhattan Project's first detonation test.

Gaseous diffusion – Process of enriching uranium by passing uranium hexafluoride through semi-permeable membranes to separate molecules of U-235 and U-238. The lighter molecules containing U-235 penetrate the barrier slightly more rapidly, and with enough stages significant separation can be accomplished.

gm – Grams.

G7 – The Group of 7, group of seven countries comprising Canada, France, Germany, Italy, Japan, the United Kingdom and the United States.

GW – Gigawatts.

^{1}H – Hydrogen.

^{2}H – Deuterium.

^{3}H – Tritium.

ha – Hectares.

Half-life – An inverse asymptotic rate in which half the isotope will decay in a given time, followed by half of the remainder in the same period again, and half of that in a further similar duration. It never reaches zero.

H-bomb – Hydrogen bomb.

HER – High Explosives Research Unit.

HEU – Highly-enriched uranium.

ICBM – Intercontinental ballistic missile.

in – Inches.

INF – Intermediate-Range Nuclear Forces.

IRBM – Intermediate-range ballistic missile.

Isotope – Any of two or more forms of the same chemical element that have the same number of protons in their nucleus or the same atomic number but have different numbers of neutrons or a different atomic mass.

k – Neutron multiplication factor.

keV – Kiloelectron Volts.

kg – Kilograms.

km – Kilometres.

kPa – Kilopascals.

kph – Kilometres per hour.

KT – Kilotonnes.

KTS – Kazakh Test Site.

kW – Kilowatts.

lb – Pounds.

Lens – Conventional explosive within an implosion-type nuclear bomb.

LEU – Low-enriched uranium.

Li^6 – Lithium isotope, source of tritium for nuclear fusion through low-energy nuclear fission.

Little Boy – Nickname of the United States' Mk 1 nuclear bomb.

LRBM – Long-range ballistic missile.

LTBT – The Limited Test Ban Treaty 1963, also known as the Partial Test Ban Treaty (PTBT).

m – Metres.

m/sec – Metres per second.

Manhattan Project – Code name for the research work that produced the United States' first nuclear bombs.

MAUD Committee – A committee established in England by Winston Churchill in 1940 to investigate the feasibility of nuclear bombs (see pages 29–31).

MD – Mass defect. The amount by which the mass of an atomic nucleus (M) is less than the total mass of that atom's electrons, neutrons and protons, and a measure of the binding energy of its nucleus.

MeV – Millions of electron volts.

mg – Milligrams.

Micronewton – Measure of force.

min – Minutes.

MIRV – Multiple independently targeted re-entry vehicles.

Mk – Mark.

MRBM – Medium-range ballistic missile.

MSBS – Mer-sol balistique stratégique (sea-to-surface ballistic missile).

MT – Megatonnes.

MW – Megawatts.

MWe – Megawatts electric.

NACA – National Advisory Committee for Aeronautics.

NATO – North Atlantic Treaty Organisation.

NDB – Nuclear depth bomb; also nuclear detonation bomb.

New START – New Strategic Arms Reduction Treaty.

NPG – Nuclear Physics Group.

NPT – Nuclear Non-Proliferation Treaty.

NSG – Nuclear Suppliers Group.

NTS – Nevada Test Site.

OR – Operational Requirement.

OSRD – Office of Scientific Research & Development.

oz – Ounces.

PAC – Penetration aid carrier.

PAEC – Pakistan Atomic Energy Commission.

PBV – Post-boost vehicle.

Penaid – Penetration aid.

Periodic table – Chemical elements ordered by atomic number and arranged in rows so that elements with similar chemical properties are grouped together.

Pile – Early name for a nuclear reactor.

Pit – Area containing the conventional charges within an implosion-type nuclear bomb.

PNE – Peaceful nuclear explosion.

PRC – People's Republic of China.

PTBT – The Partial Test Ban Treaty 1963, also known as the Limited Test Ban Treaty (LTBT).

PUREX – Plutonium-uranium redox extraction.

PX – US military commissary.

Quantum mechanics – Theory of scientific laws by which the mechanics of atoms, molecules and sub-atomic matter are governed by the principle of uncertainty, which dictates that the presence of a particle will never be predictable but only statistically probable over a protracted period.

Reynolds number – A dimensionless number that is the ratio of inertial forces to viscous forces, a fundamental parameter in fluid mechanics that governs the way flow patterns in a fluid (liquid or gas) depart from laminar flow to turbulent conditions.

RSA – Republic of South Africa.

SALT – Strategic Arms Limitation Talks.

SEATO – South East Asia Treaty Organisation.

sec – Seconds.

SEP – Société Européene de Propulsion.

SLBM – Submarine-launched ballistic missile.

SRBM – Short-range ballistic missile.

SSAW – Committee for Scientific Survey of Air Warfare.

SSBN – Ship Submersible Ballistic Nuclear.

SSBS – Sol-sol balistique stratégique (surface-to-surface strategic ballistic missile).

SSBT – Sol-sol balistique tactique (surface-to-surface tactical ballistic missile).

SSN – Nuclear-powered submarine.

START – New Strategic Arms Reduction Treaty.

Super or Superbomb – Code name for thermonuclear hydrogen bomb.

SWU – Separative work unit.

SWUs/yr – Separative work units per year.

TNT – Trinitrotoluene explosive.

Toss-bombing – A bombing method in which the aircraft delivering a bomb pulls upward as it releases its load, thereby increasing the bomb's flight time and obviating the need to fly over the target.

UF$_4$ – Uranium tetrafluoride.

UF$_6$ – Gaseous hexafluoride.

U$_3$O$_8$ – Uranium oxide.

UKAEA – United Kingdom Atomic Energy Authority.

UN – United Nations.

Uncertainty principle – See quantum mechanics.

Urchin – A fission bomb's neutron initiator.

US – United States.

USAF – United States Air Force.

USSR – Union of Soviet Socialist Republics.

UTC – Coordinated Universal Time (strictly speaking Universal Time Clock).

W – Watts.

yd – Yards.

A Poseidon C-3 Submarine Launched Ballistic Missile breaks the surface after launch from the *USS Grant*, epitomising the nuclear threat in its most stealthy form. *(US Navy)*